KW-279-341

An Introduction to
COASTAL ZONE ECONOMICS:
CONCEPTS, METHODS, AND CASE STUDIES

Steven F. Edwards

National Marine Fisheries Service
Northeast Fisheries Center

Taylor & Francis
New York • Philadelphia • London

USA   Publishing Office: Taylor & Francis · New York
3 East 44th St., New York, NY 10017

           Sales Office: Taylor & Francis · Philadelphia
242 Cherry St., Philadelphia, PA 19106-1906

UK   Taylor & Francis Ltd.
4 John St., London WC1N 2ET

An Introduction to Coastal Zone Economics

Copyright © 1987 Steven F. Edwards

*All rights reserved. No part of this publication may be reproduced, stored
in a retrieval system, or transmitted, in any form or by any means,
electronic, electrostatic, magnetic tape, mechanical, photocopying, recording
or otherwise, without the prior permission of the copyright owner.*

First published 1987
Printed in the United States of America

Library of Congress Cataloging in Publication Data

Edwards, Steven F.
   An introduction to coastal zone economics.

   Bibliography: p.
   Includes index.
   1. Coastal zone management. 2. Coastal zone
management—Case studies. 3. Coasts—Economic aspects.
4. Coasts—Economic aspects—Case studies. 5. Coasts—
Recreational use—Case Studies. I. Title. II. Title:
Coastal zone economics.
HT388.E38   1988   333.91′7   87-19431
ISBN 0-8448-1530-6 ✓

90 0647622 5

WITHDRAWN
FROM
UNIVERSITY OF PLYMOUTH
LIBRARY SERVICES

To my parents
with love and appreciation

PLYMOUTH POLYTECHNIC
LIBRARY

Accn
No   211674-1

Class
No   333.917 EDW

Cont
No   0844815306

Contents

v

Preface

If a reporter stopped you on the street and asked you to define economics and economic value, what would you say? Your responses would probably involve household and business finances and characteristics of the national economy such as prices, markets, and unemployment. At this level of inquiry, most responses would be correct in a general way. However, if the reporter pressed for precise definitions, we soon would find some fundamental disagreements between economists and non-economists. Among other things, we probably would have different definitions of costs and profits, of the relationship between price and value, and of whether markets are necessary for economic value to exist.

The purpose of this book is to promote the optimal use of coastal resources through an improved understanding of economics. The underlying premise is that economic analysis broadly construed, as opposed to narrowly confined to costs and markets, is a powerful tool for illuminating tradeoffs among conflicting allocations of scarce coastal resources. Its power rests in its simplicity and broad scope. That is, much of economic thought is reducible to a few concepts which can be applied to a wide range of problems. Although there is no field of study called "coastal zone economics" per se, the concepts that we need are available from the literatures of microeconomics, welfare economics, public finance, and environmental and resource economics.

Although the subject matter is economics, the book is written primarily for non-economists and for students enrolled in marine resource management courses. Policy makers, planners, other professionals, and concerned citizens encounter resource allocation issues daily; however, most people are not prepared to assess fully the economic content or implications of these issues. Alleged economic arguments invoked by those who are pro- or anti-economic growth are nearly always incomplete and often mistaken and misused. The reader should expect to learn enough about economics to evaluate critically such arguments. You are invited to approach the book with an open mind and to consider the time that you spend to be an effort to learn a new language that happens to use English words.

The tone of the book is intended to be objective. The emphasis that economics places on the efficient uses of scarce resources requires an awareness of all benefits—so-called non-market benefits as well as consumer satisfaction and economic profit. Although non-market benefits such as for clean water or for access to public beaches are not fully or even partially revealed in commercial markets in many cases, they have economic content nonetheless. The discussion of these benefits is necessary for a complete understanding of economics and should not be construed as a preservation bias.

Lest you are shying away from reading further, let me remind you that the book is a selective introduction to economics that focuses on coastal resources. It is not a handbook that details quantitative, analytical methods in cookbook fashion. It is a primer that should be read comfortably. Advanced concepts which are relevant to the discourse are presented in non-technical ways. The only mathematical talents which one needs are a familiarity with algebra and an ability to follow simple graphs.

My hopes are that the concepts and methods are presented clearly and that they become useful when forming coastal resource policies. Any comments about the success or failure of these hopes are welcomed.

The chapters are arranged as follows. The four chapters in Part One introduce the reader to important economic concepts and methods. Chapter 1 argues for more frequent and wider applications of economic analysis to coastal resource use problems, thereby justifying the need for the book. Chapter 2 defines and illustrates several basic economic concepts using the coastal resource, shrimp, as a pedagogical aid. Commonly used methods for the analysis of resource use are introduced in Chapter 3. These concepts and methods are inculcated in Chapter 4 where ten stereotypic arguments are critiqued.

Part Two contains six chapters covering economic case studies of Galapagos Islands tourism, relative sea level rise in Bangladesh, beach erosion, water quality in coastal lagoons, potable groundwater, and traditional economic growth. These chapters both demonstrate the application of economic analysis to coastal resource management problems and illustrate the use of several methods discussed in Part One. Chapter 11 contains parting remarks.

I am indebted to several organizations and individuals for assistance. Generous financial support was provided by the J. N. Pew Jr. Charitable Trust and the Woods Hole Oceanographic Institution's Marine Policy Center. The Department of Commerce, National Oceanic and Atmospheric Administration, National Sea Grant College Program provided additional seed money for the preparation of a precursor to this book. Dr. K. O. Emery, Scientist Emeritus in the Geology and Geophysics Department of the Woods Hole Oceanographic Institution, deserves special thanks for helping me make this book come to fruition. Dr. David Storey in the Department of Agricultural and Resource Economics, University of Massachusetts, read and provided constructive comments on the earlier draft, as did Judith Fenwick, Donna Edwards, and Elaine Edwards. Projections of future relative sea level in Bangladesh were provided by geologists John Milliman and David Aubrey of the Woods Hole Oceanographic Institution. Finally, Ethel LeFave and Jane Zentz provided able secretarial support.

Chapter 1

Economics: "It's Everywhere; It's Everywhere"

Man has occupied the world's coastal zone for centuries. Historically, entire economies prospered from industries linked to trade and renewable resources such as fisheries. Nowadays coastal economies are more diverse, including tourism and the extraction of oil and other non-renewable resources. The labor force in these sectors still comprises much of the demand for construction, food, clothing, transportation, and recreation. In turn, these industries employ even more people. There is little doubt that land, fish stocks, oil, sand and gravel, and other coastal resources will continue to be utilized to promote economic growth.

People are drawn to the sea by its environs as well as employment opportunities, however. Swimming, boating, recreational fishing, wildlife photography, and simply gazing at the ocean are just a few of our valued pastimes. Higher per capita incomes, greater amounts of leisure time, and increased mobility should continue to facilitate the pursuit of these interests well into the future. Unfortunately, their value to coastal populations is not fully or, in some cases, even partially revealed in traditional markets.

At one time, the world's coastal zone accommodated all interests in its resources—even mutually exclusive uses. Times have changed. The diverse needs and wants of a large and rapidly growing coastal population are now severely limited in many parts of the world by the fixed supply of coastal resources, by the limited capacity of natural ecosystems to assimilate man's impacts, and by legal systems of property rights that favor private ownership. Often conflict arises directly between two well functioning markets. For example, condominiums are displacing private marinas along much of the shoreline in the United States. In other cases, however, markets do not reflect the full effects of their activities. For example, pollution of estuaries with sewage and industrial wastes affects the productivity of commercial and recreational fishermen because the pollution reduces the growth and availability of fish and shellfish stocks. Other global examples of "externalities" include sewage pollution of coastal aquifers and crowding of recreational sites.

In still other cases, an inability to exclude people from a coastal resource or its benefits precludes the establishment of private markets. Non-exclusivity characterizes public beaches and access to them, as well as the many benefits that salt marshes and mangroves provide mankind, including storm protection, waste assimilation, and opportunities for bird watching. These resources tend to be developed without specific knowledge of the public benefits that are foregone.

In addition to competition for coastal resources, there are conflicts between the wills of mankind and "Mother Nature." Coastal storms, shoreline erosion, and changes in relative sea level are persistent threats to coastal properties, beaches, and groundwater quality. These threats exacerbate the use conflicts.

Overview

Whether by default or by design, our institutions allocate coastal resources among commercial and non-market uses. At the risk of oversimplifying an undoubtedly complex decision-making process, one can say that governments, conservation organizations, and citizen groups modify the directives of markets so as for account for non-market values. The underlying premise of this book is that economic analysis broadly construed, as opposed to narrowly confined to costs and markets, can play a valuable role in this decision-making process by illuminating tradeoffs among conflicting allocations of scarce coastal resources. That is, economics is ideally (but not necessarily uniquely) suited for assessing the outcomes and impacts of allocating resources among market and non-market uses.

This book tries to promote the efficient use of the coastal resources throughout the world by introducing applied economics to resource planners, government officials, lawyers, scientists, engineers, students, and concerned citizens. Although admittedly there is no field of study called "coastal zone economics," the concepts and methods that we need are available from the literatures of microeconomics, welfare economics, public finance, and environmental and resource economics. Chapters 2 and 3 describe concepts and methods which are necessary for a solid foundation in economics and explain why and how economic analysis can be used to assess a wide range of interests, including non-market values. In turn, the concepts and methods are inculcated and extended throughout the remainder of the book. Chapter 4 critiques ten familiar stereotypic arguments about economic growth. In addition, the six chapters in Part Two use case studies to illustrate in concrete ways how economic analysis can be used to assess coastal resource management issues. These case studies cover Galapagos Islands tourism (Chapter 5), relative sea level rise in Bangladesh (Chapter 6), beach erosion (Chapter 7), water quality in coastal lagoons (Chapter 8), potable groundwater (Chapter 9), and traditional economic growth on Cape Cod, Massachusetts (Chapter 10). Chapter 11 is a final pitch for using economics to study the value of coastal resources.

PART ONE: CONCEPTS AND METHODS

Chapter 2

Basic Concepts

Most economic concepts are easy to understand. Indeed, "thinking eco-nomically" is a logical process that often appeals to common sense. In this chapter, several basic concepts are defined and illustrated graphi-cally. In addition, the foundation for economic valuations is described. As you will see, neither markets, prices, consumption, nor use is a re-quirement for an economic analysis of resource value.

What is Economic Value and How is it Measured?

Most of the book is devoted to answering questions about economic value and how it is measured. We begin with a definition of economics that suits our purposes:

> Economics is a study of how people allocate scarce resources among things that provide benefits.

While other scientists such as chemists, biologists, and ecologists study the "behavior" of molecules, organisms, and natural ecosystems, econ-omists study the behavior of people. In particular, economists are spe-cifically preoccupied with behavior that allocates resources. Individuals allocate personal income and time over things that give them satisfaction. Groups of people control larger resources such as offshore oil deposits and beaches. Regardless of the size of the resources being studied, how-ever, economics is a behavioral science.

The key words in our deceptively simple definition of economics are "allocate," "scarce," and "benefits." Allocation involves choices based on goals, or objectives. Scarcity implies that there are not enough re-sources to satisfy the objectives completely. Few of us have sufficient incomes and leisure time to buy and do whatever we want. Finally, the definition specifies that the objectives involve expected benefits.

Economics (as practiced by most economists in the western world) rolls these three notions into a simple yet powerful proposition pertaining to a general objective: people allocate resources in order to *maximize* benefits net of costs. As we shall see in what follows, benefits may be personal satisfaction or profit. Thus, the economic value of something is determined entirely by its ability to generate satisfaction and profit for mankind. Economists do not decide what these benefits should be, however. In contrast, economists objectively measure the value that people at large assign to goods, services, and the environment.

It is also important to understand a second proposition that economists use: people are indifferent between things which provide the same level of satisfaction or profit. Economic value is entirely anthropocentric and self-interested. In this context, *indifference* does not imply that a person or society could care less about a particular combination of commodities. The indifference is *between* dissimilar combinations that yield the same satisfaction. Similarly, entrepreneurs are considered to be indifferent among combinations of commodities that they sell or employ which yield the same level of profit.

Willingness-to-pay, Economic Demand, and Consumer Surplus

For the time being, let's concentrate on familiar market concepts, beginning with consumers and commodities (i.e., goods and services).

According to economic theory, a person's willingness-to-pay for a commodity depends, in part, on the amount of satisfaction that he/she expects to derive from its use. (Constant use of he/she and its variations is distracting. For the sake of simplicity, I shall use masculine pronouns, hoping not to offend anyone.) Economists refer to levels of personal satisfaction as levels of personal *utility*. Although it would be extremely difficult to measure utility per se, it is possible to observe how people allocate their income across commodities and thereby infer estimates of economic value from their revealed preferences.

As a pedogogical aid to explain several important concepts, consider your consumption of shrimp *during a year*. This example will take you through the year as you buy and consume shrimp. (If you do not like shrimp, consider scallops, fish, lobsters, or a coastal resource that you buy which is physically divisible into separate units measured by weight or pieces.) We begin by asking, what is the absolute *maximum* that you would be willing to pay for the *first* pound of shrimp? Let's say it's $25 and that you actually pay this amount. Next we ask, what is the most that you would be willing to pay for the next or second pound of shrimp during the same year? Most likely, the first pound satisfied a significant part of your desire for shrimp ("We had shrimp just last month"). The most that you would be willing to pay is probably less than $25. Let's say it's $20. Again, you pay $20 for the second pound and consume it. Our exercise continues for a third pound and so on to, say, six pounds with the additional levels of satisfaction corresponding to a willingness-to-pay of $15, $10, $5, and nothing. In other words, you are satiated after consuming five pounds of shrimp in one year; the sixth pound would yield no additional satisfaction ("I'm sick of shrimp"). Economists are so confident that additional utility decreases as consumption increases that we call it a law (actually a statistical law)—the *law of diminishing marginal utility*.

The concept *marginal* is extremely important in economics, but it can be confusing because most of us are accustomed to thinking more in terms of averages. The adjective "marginal" simply refers to one additional unit at a time. In our example, *marginal utility* refers to the additional levels of satisfaction derived from consuming shrimp one pound at a time— $25, then $20, then $15, and so on. (Notice that the marginal value of the third pound is $15, whereas the *average* value of the first three pounds is $20.) The fact that the utility you get from each pound of shrimp decreases as you consume more is manifest in a decreasing willingness-to-pay for subsequent amounts. This is illustrated in Figure 2.1a, using a bar graph. Figure 2.1b illustrates the same information except that the plot is smoothed. The line in this figure is called a *demand curve*. A demand curve delineates your *marginal willingness-to-pay* for additional units of a commodity.

Next we inquire, what is the total economic value of six pounds of shrimp to you during this year? Total economic value is approximated by your total maximum willingness-to-pay, which in our example is $25 + $20 + $15 + $10 + $5 + $0 = $75. Graphically, it is the sum of the rectangles in Figure 2.1a, or, equivalently, the area under the demand curve in Figure 2.1b.

Thus far, I have not mentioned the *price* of shrimp. If a pound of shrimp actually costs $10, how many pounds would you buy this year?

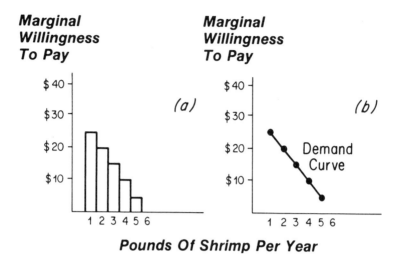

Figure 2.1 Individual demand curve.

Again in our example, the first three pounds each provide more than the equivalent of $10 of utility. However, you are just willing to pay $10 for the fourth pound; hence, *quantity demanded* is four.

Notice that the total level of satisfaction from consuming four pounds of shrimp is approximated by $70 ($25 + $20 + $15 + $10). Yet, you pay only $40 for the four pounds ($10 × 4). The remaining $30 worth of satisfaction is a net benefit since you are getting more than what you paid for. In general, *consumer surplus* is the amount of satisfaction which is in excess of payment. Graphically, consumer's surplus is measured by the area behind a demand curve but above price (Figure 2.2a). Consumers maximize utility and surplus by consuming commodities until price is equal to their marginal willingness-to-pay. Beyond that point, additional utility is less than the commodity's price, and total utility declines.

Other people will have different demand curves for shrimp. For example, the person depicted in Figure 2.2b is just willing to pay $10 for one pound of shrimp during the year. This person derives no consumer surplus from shrimp when the price is $10. Figure 2.2c depicts the demand for someone unwilling to buy even one pound of shrimp at $10. The price would have to be $5 for him to even participate in the market. Finally, Figure 2.2d depicts the demand for someone who buys six pounds of shrimp per year at $10.

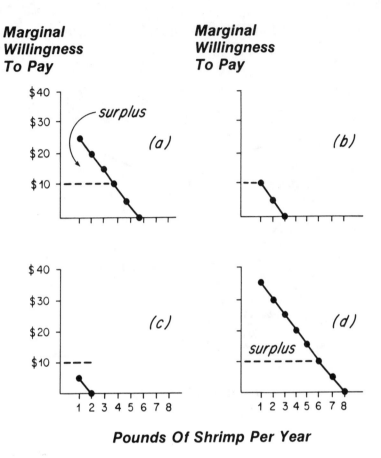

Figure 2.2 Demand curves for different individuals.

We can use the *individual demand curves* in Figure 2.2 to develop a combined, *aggregate demand curve* for the four people. At $35, the total quantity demanded by the four people is one pound (Figure 2.3). At $30, quantity demanded is two pounds and so on, until, at $5, the total quantity demanded is fifteen pounds of shrimp. At a price of $10, the four people buy 11 pounds. This quantity corresponds approximately to $215 in total satisfaction (approximated by the total area under the demand curve between one and 11 pounds), $110 in total expenditures ($10 × 11), and $105 in consumer surplus ($215 − $110 = $105). Again, notice that the total economic value of shrimp to these four people is considerably more than what most people probably would calculate it to be—price times

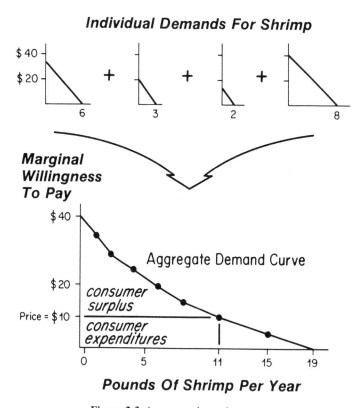

Figure 2.3 Aggregate demand curve.

quantity, or $110. The same process could be followed to develop a *market demand curve* for all consumers in the shrimp market.

There are several reasons why demand curves might differ among people. First, people might have different tastes and preferences. Thus, the individual in Figure 2.2d might like shrimp a lot more than the others (i.e., get more personal utility from consuming shrimp) and rank its use higher than for other commodities, thereby being willing to pay more for it.

Second, choices of what and how much to buy are constrained in part by income. For example, even if the person depicted in Figure 2.2c has the same tastes and preferences as the person in Figure 2.2a, he might not be able to afford to buy shrimp at $10 per pound. Hence, *price relative to income* is an important determinant of demand too. For most commodities, the higher your income, the greater your willingness-to-pay.

Finally, the prices of *substitutes* (possibly scallops or crabs) and *complements* (i.e., a commodity that goes well with a shrimp dinner—possibly wine or spaghetti) may differ among consumers. Thus, even if preferences and incomes are identical, the demand curve depicted in Figure 2.2b could be lower than in Figure 2.2a because the price of a substitute to this person is relatively low compared to shrimp, or the price of a complement is relatively high. That is, the price of a commodity *relative to the prices of substitutes and complements* is a third important consideration for consumer demand. In general, if the price of a substitute decreases (increases), your willingness-to-pay for a commodity will decrease (increase). Conversely, if the price of a complement decreases (increases), your willingness-to-pay for a commodity will increase (decrease). The point is that factors other than the price of a commodity affect willingness-to-pay and, therefore, estimates of total economic value and consumer surplus.

Consumer surplus is the most crucial concept in the measurement of economic benefits for consumers. As we have seen, it is a component of satisfaction that consumers do not pay. It is a very real phenomenon. For example, if a seafood market could charge the person in Figure 2.2a his maximum willingness-to-pay for each of the first four pounds of shrimp (this is called *price discrimination*), the customer certainly would lose an amount of satisfaction equivalent to $30. After all, the person would have $30 less to spend on other commodities that also yield utility.

Consider consumer surplus from another perspective. If the seafood market had a sale on shrimp or something that you usually buy, you would save from the lower price. The savings could buy even more shrimp or other things that give you utility. As a result, your total utility increases.

This perspective illustrates the importance of estimating *changes* in consumer surplus in economic analysis. Suppose that fishermen have record landings of shrimp this year, substantially increasing supply over last year. This could result in a price reduction for shrimp sold in retail markets to, say, $9 per pound. What effect would this have on consumers?

Figure 2.4, which illustrates a hypothetical market demand curve for all consumers, will be used to answer this question. At $10, let's say that 600 million pounds of shrimp are bought per year from seafood markets. To say that the benefit of the price reduction for consumers is $600 mil-

**Price
Per Pound**

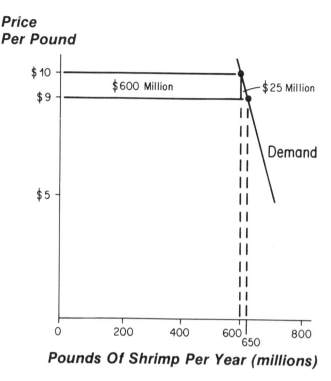

Pounds Of Shrimp Per Year (millions)

Figure 2.4 Market demand and gain in consumer surplus due to a price reduction.

lion [i.e., 600 million × ($10 − $9)] would underestimate the actual gain in surplus by $25 million ($25 million is the value of the small triangular area in Figure 2.4). That is, the 10% reduction in price increased quantity demanded by more than 8%, resulting in a gain in satisfaction that is equivalent to $600 million for the original 600 million pounds of shrimp, plus $25 million associated with the additional 50 million pounds.

Another way to look at this gain in surplus is to consider the effect on a person whose demand for shrimp is that in Figure 2.5. Here we use consumer surplus to explain further the concept of indifference. If this person's annual income is $25,000 after taxes, he would be indifferent between (a) 4 pounds of shrimp and commodities bought with $25,000 − (4 × $10) = $24,960 and (b) 6 pounds of shrimp and commodities bought with $25,000 − (6 × $9) − [($10 − $9) × 4 + (1/2) × ($10 − $9) × (6 − 4)] = $24,941 where the expression in brackets is the gain

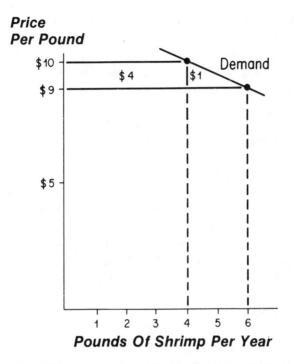

Figure 2.5 Individual demand and gain in consumer surplus due to a price reduction.

in consumer surplus from the four original pounds plus the surplus on the fifth and sixth pounds. In other words, the person is indifferent between initial conditions where price is $10 and the final condition where price has declined but income is reduced by the increased surplus. You might want to review this example.

You have been exposed to a lot of new information. Let's summarize the salient points. First, economic value can be expressed monetarily simply because you and I allocate our incomes over the large number of commodities which yield varying amounts of personal satisfaction. Second, price is related to value because it reveals what people are willing to pay for a commodity. More precisely, price reveals the marginal value corresponding to the last unit that someone purchases. Third, if someone claims that the value of something is price times quantity demanded, you know that this underestimates total economic value by consumer surplus. Thus, consumer surplus is the net value, or net benefit that consumers derive from the commodity, and is equal to the difference between total willingness-to-pay and actual expenditures. Finally, *changes* in consumer surplus represent changes in net benefits. Economists call the analysis of changes in surpluses *welfare analysis* since changes in personal welfare or well-being are being assessed.

Marginal Cost and Economic Profit

We switch now to concepts related to the production and sale of commodities in markets. Using the seafood market as an example of a company, or *firm,* what would economists consider to be costs in the sale of shrimp? Certainly, the owner of the market has to pay for shrimp that he sells, for labor, for electricity, and possibly for a mortgage on the capital (e.g., building, refrigeration units) and land. Economists call these *factors of production,* or *inputs.* They can be classified as *fixed factors* (e.g., the building), the use of which does not change in the *short run* by definition regardless of the amount of shrimp that is sold, and *variable factors* (e.g., shrimp, labor) which can be varied at any time (in the *long run,* all factors of production are variable by definition).

The firm's accountant might stop here to determine total costs. However, economists insist on including the costs associated with the entrepreneur's time and investment as well. After all, the owner of the market could earn an income, for example, by selling the market and investing into a seafood restaurant. However, his best opportunity is foregone by choosing to operate the seafood market. The foregone value of one's highest paying alternative is called an *opportunity cost* (more on this below).

Now we can define *economic profit* as economists see it. Economic profit is total revenue minus total economic costs, including the opportunity cost of the owner's investment and time. Let's assume that the firm's cost of selling each additional pound of shrimp during the year is constant regardless of the number of pounds sold. If cost is $6 per pound but the firm receives $10 per pound, it makes an economic profit of $4 per pound. If the entrepreneur sells 400 pounds per year, total economic profit is $1,600. As for consumer surplus, economic profit is a net benefit.

Let's consider the cost relationship in greater detail. Typically, the additional, *marginal cost* of producing additional units of a commodity increases as the number of units sold or produced increases. This is a technical matter but can be explained using our shrimp example if the *productivity* of labor decreases as output increases. Productivity refers to the amount of shrimp that one worker can sell. For example, as more labor is hired to sell shrimp, employees will get into each other's way at some point, particularly at the cash register. For this reason, an additional worker can not sell as much shrimp as the one hired previously. Furthermore, the new worker also reduces the amount of shrimp that all other workers can sell. Thus, to double the sale of shrimp, the owner would have to more than double the number of employees. As a result, the marginal cost of selling shrimp increases as sales increase. This is illustrated with a bar graph in Figure 2.6a where the total cost of variable inputs is determined by adding the value of the bars. As we did for de-

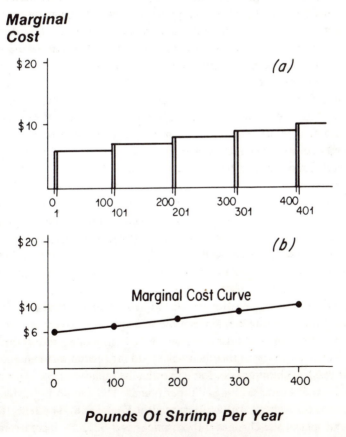

Figure 2.6 Marginal cost curve for a firm.

mand, the bar graph can be smoothed with a line to approximate the marginal cost curve (Figure 2.6b). The area below the line is the total economic cost of variable inputs such as shrimp and labor (additional cost concepts such as fixed costs are discussed in Chapter 4).

In a competitive industry (defined below) where there are many seafood markets each selling shrimp at the same price, a single market will sell or supply shrimp until the cost of providing an additional pound—the marginal cost—is just equal to the corresponding amount of additional revenue. In this case, *marginal revenue* is constant and equal to price since the owner receives the same revenue for each pound until marginal revenue equals marginal cost, and, therefore, *marginal profit* is zero. That is, an entrepreneur in a competitive industry maximizes profit by producing output until the additional revenue from the sale of the last unit equals the additional cost of producing it. Beyond this point, marginal cost is greater than marginal revenue, and total profit actually declines. Since the competitive firm's marginal cost curve delineates the *quantity supplied* at each price, it is also the firm's *supply curve*.

These concepts are illustrated in Figure 2.7. Total economic profit is the sum of marginal profits on each pound of shrimp. Graphically, it is the area above the marginal cost curve and below price. And as we did for individual demand curves, we can add individual supply curves to get an aggregate supply curve as in Figure 2.7.

Exactly who receives the profit is also of interest. Strictly speaking, the owner of the fixed assets (the building and land), who we have assumed to be the manager, keeps the profit. If the manager is actually renting the seafood market, the owner could increase the leasee's rent until he captures all the economic profit, which, you recall, excludes the leasee's opportunity cost. At this point, the leasee is indifferent between managing the seafood market and working at his next best alternative since either job would provide the same income.

Finally, consider the effect of the $1 reduction in retail price on the owner's profits (assume again that the manager and owner are the same person). However, we must also consider the effect of a reduction in the wholesale or input price that the owner paid for shrimp that resulted from the increase in shrimp catch. Let's say that the input price for shrimp paid by the owner decreases by $1.50 per pound. This would lower the marginal cost curve by $1.50 as illustrated in Figure 2.8. The net effect in this example of the changes in both prices is a $212.50 increase in economic profit since the loss due to the reduction in the retail price (area

Individual Supply Curves

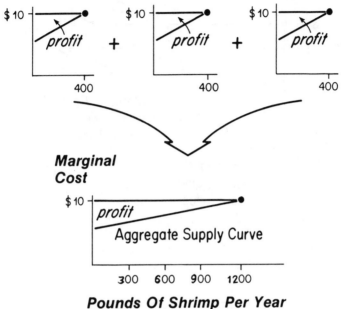

Figure 2.7 Aggregate supply curve.

ABCD, or $350) is less than the gain due to the reduction in costs (area DEFG, or $562.50). Notice that if someone estimated the change in profit by either multiplying the change in input price by the original number of pounds of shrimp sold (gain of $1.50 × 400 = $600) or by multiplying the change in the retail price of shrimp by the original number of pounds (loss of $1 × 400 = $400), he would over- and under-estimate, respectively, the change in profit (also notice that, by virtue of a downward sloping demand curve, the seafood market does not pass its entire cost savings to consumers).

In summary, economic profit is a measure of the net benefits received by owners of factors of production—the owner of the seafood market, the partners in a marina, a large tourism corporation. It differs significantly from the commonplace use of the word "profit" because the value of foregone earnings is a real cost—an opportunity cost. We will return to opportunity costs soon.

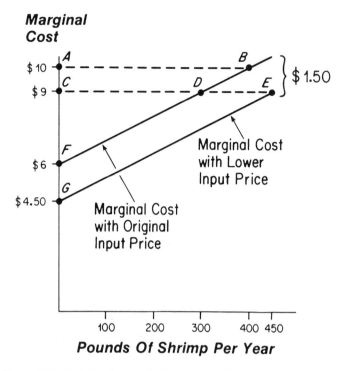

Figure 2.8 Effect of reductions in input price and output price on a firm's marginal cost and profit.

A Few Comments About Markets

A few comments about markets are necessary to explain how prices emerge from economic behavior and to set the stage for the discussion of market failures and inefficiencies. A *market* for a commodity consists of a group of potential buyers and sellers. Under competitive conditions, the actions of all buyers and sellers determine market price. Thus, markets and prices emerge from economic behavior. If sellers set the price too high, they will not sell their entire inventory. If the price is set too low, they will run out of their product too soon. At equilibrium, market price is such that consumers are just willing to pay the price for all units available and suppliers are just willing to sell the same amount at that price. This is illustrated in Figure 2.9 where demand and supply intersect at $10. This price *clears the market*. Thus, the interaction of market demand and supply determines the price.

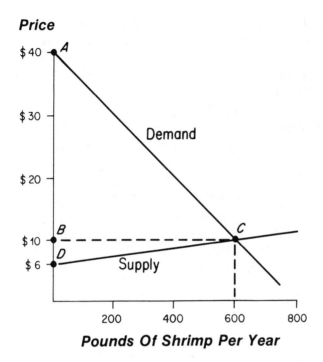

Figure 2.9 Demand, supply, and price in a competitive market.

The concept of a competitive market holds a special place in economists' hearts. In essence, pure competition implies (among other things) that (1) there are numerous firms and consumers who *individually* have no effect on prices and (2) that all costs and benefits are captured by markets. Under these conditions, social benefits as measured by economists (i.e., consumer surplus plus producer profits) are maximized. This is shown in Figure 2.9 for a single market. At a price of $10, the marginal benefits to society (i.e., all consumers) are just equal to the marginal opportunity costs to society (i.e., the marginal opportunity costs of factors used in the sale of shrimp). Consumer surplus in this market, summed across all buyers, is area ABC. Similarly, economic profit for all producers is area BCD. If the price of shrimp is set higher or lower (and this is not due to a shift in demand or supply caused by changes in other prices, in income, or in the number of consumers and firms), total social benefits would be reduced.

The economic impacts of the price reductions on all consumers and firms can be illustrated with market demand and supply curves (Figure 2.10). As we saw in Figure 2.4, the $1 reduction in retail price results in a gain in consumer surplus equal to $625 million (area BCFD). In addition, profits in seafood markets increase by area DFH minus area BCG, or $262.5 million. The net effect on the industry is equivalent to area GCFH, or a gain of $887.5 million.

We shall return to the analysis of changes in economic benefits in Chapter 3. Before leaving our discussion of markets, though, I want to plant another seed. You might be wondering what the change in the seafood market's costs means to groups other than consumers. Recall that the marginal cost curve delineates the total cost of using and employing *all* variable factors of production—shrimp, labor, electricity, and management in our example. Looked at from another perspective, it is the income earned by the owners of the productive inputs which are used to sell shrimp. In our example, fishermen are getting a lower price for their catch. Labor is also

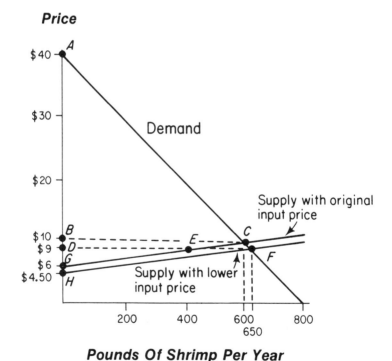

Pounds Of Shrimp Per Year

Figure 2.10 Effects of reductions in input price and output price on market supply, consumer surplus, and profit.

a variable input, however. We own our skills and sell our time in labor markets at the prevailing wage rate. The way that changes in income are treated in economic analyses is an important theoretical and practical matter that will be considered at the end of Chapter 3. The important point for now is that a change in price in one market might have welfare effects in other markets too.

It is of considerable interest that many individuals acting in their own self-interest could actually maximize the social benefits of commodities. Under perfectly competitive conditions, prices allocate inputs and commodities to the most efficient markets. By *efficiency,* I refer to an optimal allocation of inputs in the production of commodities which yield maximum net economic benefits for society. Of course, the world is not set up this way. All markets are not perfectly competitive. Indeed, many markets have monopolistic elements. In addition, markets often fail to internalize all costs and benefits associated with their activities. Furthermore, there are many important cases when markets do not exist at all, and the preferences of people are not as easily determined. These exceptions characterize most if not all resource conflicts in the coastal zone. We shall return to these points soon.

Personal Discounting

Consumers and firms often make decisions that affect consumption and earnings over time. How does economics treat such decisions? Again, the answer involves indifference.

We start with a traditional example. You probably agree that it is better to receive $100 today than in a year from today because, among other things, you could deposit the money in a bank today and after one year have more than $100. If the annual interest rate is 6%, you would have $106 at the end of a year.

Let's evaluate this example from a different perspective and ask what amount of money would make you indifferent between receiving that amount in one year versus $100 today? If your answer is $106, you are indifferent between the satisfaction that $100 could buy you today and that $106 could buy you in one year. This is equivalent to saying that you *discount* future consumption by 6%. Hence, the *present value* of $106 worth of commodities in one year is $100 today.

This relationship between present and future values is expressed algebraically by the equation:

$$PV_0 = A_t/(1 + d)^t$$

where PV_0 is present value of consumption in the current year, A_t is the future value of consumption in year t, and d is your personal discount rate. This equation can be transformed algebraically to show the relationship between the discount rate and present and future values:

$$d = (A_t/PV_0)^{1/t} - 1$$

In our example of indifference between $100 today and $106 in one year (t = 1), PV_0 = $100, and $A_{t=1}$ = $106. Hence,

$$d = (\$106/\$100)^1 - 1 = 0.06, \text{ or, } 6\%$$

In general, the present value of a flow of future income is equal to

$$A_0/(1 + d)^0 + A_1/(1 + d)^1 + A_2/(1 + d)^2 + \ldots + A_T/(1 + d)^T, \text{ or,}$$

$$\sum_{t=0}^{t=T} A_t/(1 + d)^t$$

where T is the last, or terminal time period, and Σ is a symbol representing the summation over time. Thus, any future benefit, cost, or net benefit (i.e., benefit minus cost) can be standardized in terms of present values. How convenient.

Notice that the future value in the above example (i.e., $106) is expressed in *real* terms. That is, no adjustment for inflation was made. In contrast, *nominal* values include increases due to inflation over the time period(s) being considered. When projecting future prices, costs, and benefits, it is best not to add a factor for inflation since this is arbitrary and would have to be accommodated anyway by increasing the discount rate.

Discounting is an important concept for economic analysis because it organizes time series of benefits and costs in a logical and theoretically sound way and in a form which facilitates the comparison of alternatives for decision-making. For example, Table 2.1 summarizes the costs over time of buying a canoe plus accessories today for $800 versus renting one for $60 each year for vacations. Let 15 years be the life expectancy of a new canoe at which time you could salvage it for $120. Also, your maintenance costs for repairing the canoe and replacing equipment are expected to be $30 every four years beginning the fifth year. Assuming that you derive the same satisfaction from canoeing whether you rent or own, the benefits cancel across alternatives, and we are only looking for the least expensive alternative. Is it less expensive to buy a canoe or to continue renting one each year?

Using a 10% rate of personal discount, the present value of the costs of purchasing and maintaining the canoe and equipment is $844.06 (Table 2.1). However, the present value of the *salvage value* of the canoe ($28.73) must be deducted from this amount, leaving $815.33. This is considerably more than the present value of the costs of renting ($516.37). Consequently, renting is cost-effective. It is very important to notice that this

Table 2.1

Cost-effectiveness Analysis of Canoeing: What is the Cost-effective Option for Canoeing? Present Value, $PV = A_t/(1 + 0.10)^t$.

	Alternative			
	Rent ($)		Buy ($)	
t (year)	Cost	PV	Cost	PV
0	60	60.00	800	800.00
1	60	54.55	0	0
2	60	49.59	0	0
3	60	45.08	0	0
4	60	40.98	30	20.50
5	60	37.26	0	0
6	60	33.87	0	0
7	60	30.79	0	0
8	60	27.99	30	14.00
9	60	25.44	0	0
10	60	23.13	0	0
11	60	21.03	0	0
12	60	19.12	30	9.56
13	60	17.38	0	0
14	60	15.80	0	0
15	60	14.36	−120 (salvage value)	−28.73
Present Value		$471		$815

conclusion is the opposite that would be reached if you neglected to discount future costs and simply added the future values: $960 ($60 × 16) for renting versus $770 ($800 + $90 − $120) for buying.

You might wonder, however, what the costs would have to be in order for buying to be cost-effective. If the canoe and equipment were on sale for about $501 ($516.37 − $20.50 − $14.00 − $9.56 + $28.73 = $501.04) and your personal discount rate was 10%, you would be indifferent between renting and buying.

Before leaving this section, notice the effect of the discount rate on future values. Discounting quickly reduces future values to smaller and smaller present values as time extends into the future. Even at 2% discounting, the present value of $100 received in 50 years is only $37.15 (Table 2.2); in 100 years, it is $13.80. The higher the discount rate, the faster the reduction in present value.

Economic Costs are Opportunity Costs

Most people think of costs in terms of financial outlays or what they expend on inputs and commodities. Economists have a different perspective on the matter, though. As indicated above when discussing economic profit, economists consider all costs to be the value of what is given up by doing one thing instead of something else—the value of foregone opportunities, hence, opportunity costs. Thus, there is a cost associated with everything that you do. This is why economists insist that

Table 2.2
Present Value (PV) of $100 Received at the End of "t" Years. Discount rates are percentages converted to fractions (e.g., 2% is 0.02).

t (year)	Discount Rate				
	0.00	0.02	0.05	0.10	0.25
0	100	100.00	100.00	100.00	100.00
5	100	90.57	78.35	62.09	32.77
10	100	82.03	61.39	38.55	10.74
25	100	60.95	29.53	9.23	0.38
50	100	37.15	8.72	0.85	0.00
100	100	13.80	0.76	0.00	0.00

nothing is free, particularly nowadays with man's demonstrated ability to pollute all regions of the world.

Opportunity costs are related closely to the scarcity of resources. By "scarce," I mean that there is not enough of a resource to satiate all interests in it; hence, it must be allocated to different uses. Prices mediate the allocation of marketed commodities.

For the vast majority of us, income and time are scarce resources that we allocate among commodities and between work and leisure. For example, if you go sports fishing rather than swimming, the opportunity cost of that decision is the surplus foregone by not swimming (presumably, the opportunity cost is less than the surplus that you expect from fishing; otherwise, you would make the opposite decision). As another example, a decision not to work on weekends at a part-time job suggests that your leisure time is more valuable than what could be bought with the extra income. Nevertheless, the foregone income is an opportunity cost of not working. As a final example (and at the risk of giving you an excuse to stop reading), your time spent with this primer has an opportunity cost equal to the value of what you would otherwise be doing.

This thought process can be extended to firms as well. We already discussed the opportunity costs of an entrepreneur's investment in relation to the seafood market example. Owners of fixed factors of production consider the income they forego (including profit) by not selling their assets and investing them elsewhere. In addition, input prices in competitive markets reveal the marginal value of forgone production.

Opportunity costs can be considered from yet another social point of view. The question becomes, what does society as a whole lose when labor, capital, land, and the natural environment are used to produce one thing rather than another—condominiums replacing marinas, golf courses replacing wetlands, private homes replacing public beaches? Having covered opportunity costs in the context of traditional markets, it is time to consider this question within the context of unpriced resources.

Economics of Unpriced Resources

Thus far, we concentrated on concepts associated with single-goods markets—demand and supply, prices, consumption, consumer surplus, economic profit. Indeed, demand and supply are necessary for the emergence

of markets, prices, and economic profits. However, demand and supply
do not guarantee the emergence of markets and prices. To complicate matters further, economic demands do not even require consumption or use;
therefore, it is best to discuss surplus rather than "consumer" surplus. To
clarify these points, we discuss commodities which are not exchanged in
single-goods markets but which nonetheless generate economic value.

Non-exclusive Resources and Externalities

Markets emerge from special circumstances when individuals are willing
to pay for a commodity or environmental resource, when a technology
facilitates production and supply at costs less than or equal to revenues,
and, importantly, when consumers and firms can exclude others from the
benefits of their consumption and output. When there is no demand (Figure 2.11a) or when it is small compared to the marginal costs of provision
(Figure 2.11b), firms cannot cover their opportunity costs, and markets
and prices do not emerge. And without exclusivity, there is no supply
regardless of what people are willing to pay since it would not be feasible
to price the output in order to cover costs (Figure 2.11c). We concern
ourselves here with exclusivity and with factors that affect the provision
and values of environmental resources. It helps to classify coastal resources along two dimensions. First, *exclusivity* involves cultural prec-
edents and properties of the resource. For example, governments might
consider it unconstitutional to exclude the public from beaches. In addition, the migratory habits of some species of fish make it impossible
to establish individual property rights that would exclude other firms and
the public from the resource. Second, *divisibility* in final use concerns
whether the resource can be subdivided such that each individual who is
willing to pay for it can exclude all others from its benefits.

Coastal resources which are allocated in private markets are always
exclusive. For example, the seafood markets in our previous example can
exclude anyone who is unwilling to pay the market price. In addition,
the shrimp are divisible such that no one gets utility from the shrimp that
someone else buys. In this case, price rations the shrimp among consumers, and the firms earn enough revenue to cover their costs.

Next consider *open access* or *common property* resources. For example, no one owns the shrimp in their native habitat. Hence, shrimp and
most other mobile or migratory species of invertebrates and fish are non-

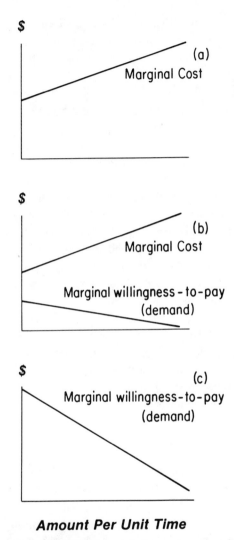

Amount Per Unit Time

Figure 2.11 Situations when markets and prices do not emerge: (a) no demand; (b) demand small relative to costs of supply; and (c) no supply due to non-exclusive property rights.

exclusive resources. However, their divisibility facilitates ownership once they are harvested. Ground water is an open access resource too, characterized by non-exclusive property rights and divisibility in use.

Conversely, there are a number of coastal resources that are exclusive and, therefore, amenable to pricing, but which are indivisible. For example, in many parts of the world, access to bays and estuaries, dive sites, and endangered whales is provided by firms although use of these environments and of whales is indivisible since one person's use does not preclude use by others. In many other cases, governments somewhat privatize coastal resources, such as state beaches, by charging entrance fees. Indeed, many communities exclude non-residents from local beaches through various parking and pricing mechanisms. Regardless, a public beach is not partitioned such that my use precludes yours (strictly speaking, my use of a specific spot precludes yours, but this is temporary).

Finally, other coastal resources are both non-exclusive and indivisible. Sometimes referred to as pure "public goods," these include scenic views, clean air and water in a general sense, the survival of endangered species, and recreational sites where access and legal uses are uncontrolled.

Allocational matters aside, next consider factors which affect the quality of coastal resources and their economic value. Coastal resources benefit mankind in three general ways. First, they provide necessary inputs into the production of marketed commodities. The examples are numerous: energy is derived from oil and gas deposits; wetlands provide habitat for the early stages of valued fish stocks; groundwater is extracted for residential, commercial, and industrial uses; offshore sand deposits are mined for construction materials; mangroves are harvested for lumber.

Second, natural environments can protect private property and our health. For example, mangroves, salt marshes, and dunes protect private property from storms. In addition, wetlands sequester pollutants, preventing them from entering into estuarine and marine food webs.

Third, certain coastal resources affect personal utility directly. Many outdoor recreational opportunities are not mediated by markets: swimming, boating, fishing, hiking, sightseeing, and wildlife photography. In these cases, we derive benefits directly from the environment rather than from markets that use coastal resources as raw materials in a production process.

Our use of coastal resources can affect their quality, however, because of the limited ability of natural and social systems to assimilate the stresses imposed by collective use. In particular, *externalities* arise when the behavior of an individual or firm affects the utility or productivity of others and the effect is involuntary and uncompensated. Although externalities can be beneficial, such as when a philanthropist donates a beach for public use, many are deleterious. Certainly, air and water pollution are negative externalities. The pollution of estuaries with sewage, heavy metals, and polychlorinated biphenols has reduced the productivity of commercial and recreational fisheries by reducing the growth of fish stocks and by causing bans on shellfish beds. The damages resulting from such activities are not internalized by markets. This is due in part to poorly defined property rights. Who, in fact, owns a public resource? Everybody, but nobody does, hence, the "problem of the commons." Also, who is culpable for pollution? The polluters or those who demand their output?

Pollution externalities are also due to a lack of exclusive property rights and to indivisibility of pollution and cleanup. For example, there is little incentive for a jewelry firm to increase costs by reducing heavy metal contamination of an estuary if it cannot exclude contiguous property owners and all recreationists from the benefits. In addition, the indivisibility of a clean estuary works against property owners paying for cleanup costs when other recreationists cannot be excluded.

In addition to reductions in environmental quality, there are externalities associated with reductions in the quantity of a coastal resource. For example, the various benefits provided by wetlands are not adequately appraised when residential development and preservation are compared. There are damages in terms of foregone benefits associated with storm protection, waste assimilation, and wildlife habitat which are related directly to the amount of wetlands. Wetlands mitigation policies, or the replacement, enhancement, or restoration of wetlands, do not take these damages into account adequately at this time.

By no means are all externalities due to market inefficiencies, however. Many recreational uses of coastal resources are mutually exclusive and do not account for their external effects on others. Thus, wind boarding, surfing, and swimming are incompatible uses of the same area of water. Duck hunting and bird watching are also incompatible if practiced simultaneously. The collective pollution of aquifers and shellfish beds with sewage from septic systems is another common example of a non-market externality.

Often, individuals' use of the environment reduces the utilities of others engaged in the *same* use. The growing demand for recreational resources along the coast is apparent at crowded beaches, fishing sites, and boating areas. Depending on one's preferences for crowds, congestion can reduce utility directly. In other cases, intertidal and shallow subtidal shellfish stocks are overfished, thereby reducing future recruitment to the recreational shellfishery.

Three concepts have been glossed over in this brief introduction to environmental and resources economics. First, it helps to distinguish between damages and costs, although the distinction can be semantical at times. *Damages* refer generally to losses in benefits (i.e., utility and profit) associated with an externality, whereas *costs* refer to the opportunity costs of resources used to produce a market or non-market commodity.

Second, we have focussed on *technical externalities*. In terms of our discussion, technical externalities affect either personal utility or the productivity of inputs and, therefore, marginal costs to producers. In contrast, *pecuniary externalities* concern indirect effects in other well-functioning markets. For example, the rapidly growing demand for property with water frontage is raising property values and the tax liability of present waterfront owners. This favors a transfer of shoreline use from "Ma and Pa" marinas to condominiums. While there may be ethical reasons for "protecting" marinas from the condominium market, considerations about what *should* be done to militate against an efficient allocation of the shoreline are not economic considerations. (Of course, potentially significant non-market benefits associated with marinas are within the purview of economics.) Whether pecuniary externalities have efficiency implications depends on employment conditions. We return to this possibility in Chapter 3.

Finally, whether an externality is *relevant* from an economics point of view depends on whether a net benefit would result from a reallocation of coastal resources among conflicting uses. When this is the case, there is an opportunity to improve efficiency by reducing the external diseconomy. If the net benefit is expected to be zero or negative, the externality is irrelevant from an economics perspective.

An Expanded Value Typology

As we just saw, market prices do not exist for non-exclusive resources or externalities. But does this mean that panoramic views of the ocean, public access to beaches, clean air and water, the preservation of whales, and so on do not generate economic value? The question is meant to underscore the commonly held misconception that markets and prices are necessary conditions for economic value. However, is it appropriate to consider the value of unpriced resources in monetary terms? To answer this question, we return to the foundations of economic thought.

Recall the introduction to this chapter. Economic value requires only that a person determines value rationally on the basis of satisfying personal wants (utility or profit) and that the person's preferences for environmental resources and market commodities are characterized by indifference. In this case, indifference implies trade-offs between environmental resources and income (actually, commodities that one can buy with disposable income) which yield the *same* level of utility. For example, someone may be indifferent between the pairs of situations depicted in Figure 2.12. In other words, he is indifferent between (a) canoeing twice a year along an estuary and the commodities that can be bought with a disposable income of $24,000, and (b) canoeing 10 times a year and a disposable income of $21,900 for other commodities. He is also indifferent between these situations and not canoeing at all, but having $28,000 to spend on other commodities.

The curve illustrated in Figure 2.12 is an *indifference curve*. These hypothetical combinations of canoeing trips and "other commodities" provide equal levels of personal utility; the person is indifferent among combinations along this curve. Of course, the pair that one actually chooses depends on a number of things that we discussed above, particularly personal income, the costs of traveling to canoeing sites, the prices of substitutes, the prices of complements (perhaps the price of a six-pack of beer or of camping equipment), and the amount of leisure time.

I used this example specifically to argue that neither physical consumption nor markets are necessary for economic value to be well-defined. For our purposes, we can divide environmental resource values into three types: *consumptive use* values, *non-consumptive use* values, and *non-use* values. Consumptive uses involve the physical use of the environment. Fishing and duck hunting are good examples. When future use is uncertain, economists refer to *option value* as the extra amount (an insurance premium, if you will) that a person is willing to pay to eliminate the risk of a future opportunity not being available. Even if you are uncertain about whether you ever will go fishing for striped bass, you might value the option to protect its breeding habitat. This amount is in addition to the *expected value* of future use. (Here, expected value refers to the utility from future use times the probability that you will actually use the resource.) This is distinguished from present value since even expected values are discounted to a present value in an economic analysis of future benefits. *Option price* is defined as the expected value of future consumption plus option value.

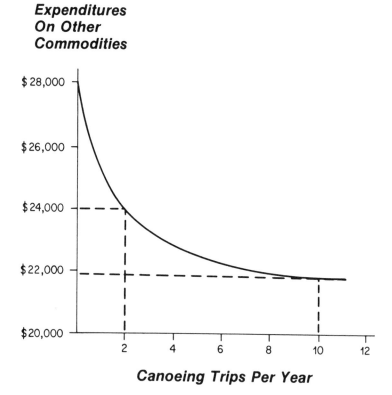

Figure 2.12 An indifference curve.

Non-consumptive use involves being at the site but not actually harvesting any resource. There are many possible examples of non-consumptive uses of coastal resources including canoeing, swimming, photography and SCUBA diving. In a sense, the experience itself, as opposed to the physical object, is being consumed. Option value and option price are defined for future, non-consumptive uses also.

Finally, non-use values, or what economists call *existence values*, concern one's maximum willingness-to-pay to preserve the environment for use by future generations (*bequest value*) and for wildlife (*preservation value*). These concepts are well-defined for those who base their maximum willingness-to-pay on the *personal satisfaction* that they receive from making bequests or preserving wildlife and whose preference structures for income and these goals are mapped by indifference curves.

"Shadow" Value

Accepting, if you will, that economic values for environmental resources are well-defined, how does one go about measuring economic value when markets do not exist? In order to answer this question—which we will do in the next chapter—one must first understand the concept *shadow value*.

Recall the helpful role that prices play in economic analyses. Prices reveal the value of the marginal unit. That is, they reveal the satisfaction of the last unit manifest in willingness-to-pay. If you had data on the amounts exchanged at different prices, you could begin to identify demand and supply curves and, therefore, to estimate consumer surplus and producers' profit.

When prices do not exist, however, you must uncover other proxies for willingness-to-pay. Such proxies are referred to as shadow prices or values. They may be for access to ocean beaches, environmental resources that are sold jointly with other commodities, such as when you buy a house with a water view, or adjustments to force markets to internalize externalities. Sometimes it is possible to discern shadow prices for small (i.e., marginal) changes in environmental resources. At other times, however, it is not feasible to subdivide the environment into small units that would correspond to marginal willingness-to-pay. For example, it makes no sense to think of one-one thousandth of a water view since it is indivisible. As a further example, it is conceptually difficult to think of making small improvements in water quality when we are really trying to value recreational opportunities. In these cases, we try to assess the entire economic value directly in terms of total willingness-to-pay. A few methods that can be used for this purpose are discussed in the next chapter.

Chapter 3

Basic Methods

This chapter outlines several methods that economists use to quantify and to compare economic values. Benefit-cost analysis is treated in greatest detail. Other methods are discussed briefly, leaving the interested reader to pursue more detailed discussions.

For our purpose, economic methods can be divided into descriptive methods which quantify demand and supply and associated economic benefits and comparative methods which compare the outcomes of alternative policy options. The comparative methods require information derived from descriptive studies.

Descriptive Methods

Descriptive methods can be divided further into traditional methods used to estimate market relationships and those used to estimate shadow values. The latter approaches are either experimental or market-related in the sense that some market data are used.

Market Methods

Data on prices, quantities sold, income, prices of substitutes, prices of inputs, etc., over time (time series data) or across markets (cross-sectional data) are used to estimate demand and supply relationships. For example, the data and corresponding graph in Figure 3.1 for pairs of annual visits and entry fees to a state beach are summarized by the equation

$$P = \$14 - (2 \times \text{Visits})$$

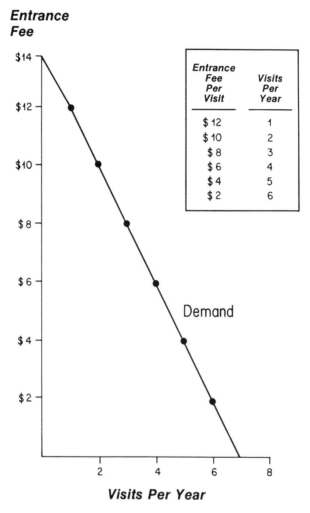

Figure 3.1 Individual demand for a state beach.

or, after transforming the equation algebraically,

$$\text{Visits} = 7 - (0.5 \times P)$$

where P is the entry price. If the entry price is $10, this person would go to the beach twice a year (7 − 0.5 × $10 = 2). Strictly speaking, the transformed equation is the demand curve, while the first equation is called an *inverse demand curve*. Notice that the value of the intercept for the inverse demand curve (i.e., 14) corresponds to where the curve hits the price axis in Figure 3.1. Also, the slope of the linear inverse demand curve (i.e., -2) is equal to the change in price divided by the change in visits between two points [e.g., $(10 - 6)/(2 - 4) = -2$].

Figure 3.2 illustrates a more detailed picture of the same demand curve that includes the effects of income and of the price of a possible substitute such as fishing on a charter boat. In this case, the inverse demand curve implied by these data is

$$P = \$12 - (2 \times \text{Visits}) + (0.00005 \times \text{Income}) + (0.025 \times P_{\text{fishing}})$$

or

$$\text{Visits} = 6 - (0.5 \times P) + (0.000025 \times \text{Income}) + (0.0125 \times P_{\text{fishing}})$$

To see how this equation is used to predict the number of visits, let $P = \$10$, Income $= \$20,000$, and $P_{\text{fishing}} = \$40$. As before, the individual would visit the beach twice a year if the entrance price is \$10 [$6 + (0.000025 \times 20,000) + (0.0125 \times 40) - (0.5 \times 10) = 2$].

In either case, but particularly in the second, multivariate case, econometrics is used to estimate statistically the numerical coefficients on each determinant of demand (including the intercept) and thereby quantify the effects of prices and income on quantity demanded. For example, every \$2 decrease (increase) in the entrance price increases (decreases) annual visits by 1 (0.5 visits $\times$ \$2), as long as Income and P_{fishing} are held constant. Therefore, if P decreased from \$10 to \$2 per visit, the number of visits would increase by four. Similarly, every \$20,000 change in income changes the number of visits by one.

Suffice it to say that a similar approach is taken to estimate supply curves and *production functions,* another important technical relationship in economics. Production functions are equations that quantify the relationship between quantities of a commodity being produced and the quantities of inputs (capital, labor, land) used in their production, given a type of technology. Whereas supply curves (like demand curves) are behavioral relationships showing how people respond when price and other factors change, production functions express physical relationships between factors of production and the amount of a commodity that is produced.

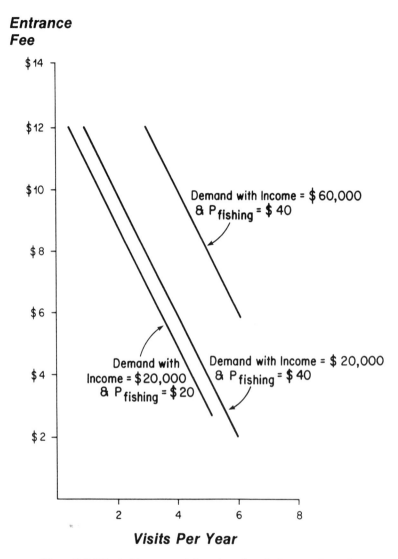

Entrance Fee

Figure 3.2 Effect of income and the price of a substitute on demand.

Nevertheless, a firm's production function is an important determinant of the supply relationship (recall the discussion of how labor productivity affects costs). Two additional functions that take advantage of this relationship are called *damage functions* and *cost functions*. Damage functions quantify the physical relationship between externalities and productivity and value the marginal change in productivity at the market price of a commodity. In contrast, cost functions are specified in terms of relative prices. The cost function approach is preferred since it can be used to derive supply curves directly and to estimate losses associated with large changes in quantities. See Freeman (1979) for a further discussion.

It is tempting to go on about various estimation techniques, potential statistical problems, the choice of model *specification* (i.e., what variables to include) and of *functional form* (e.g., linear in the variables, as above, logarithmic, or others), and alternatives to econometrics, but I shall refrain. If you have a basic understanding of the above examples, you understand the essence of an important set of analytical methods.

Market-Related Methods

Often there is no market or an insufficient amount of market data to allow the direct estimation of demand and supply for environmental resources. However, there are cases when peoples' preferences are revealed indirectly in related markets. Many non-exclusive coastal resources and externalities can be studied using one of the three following market-related techniques.

The first technique takes advantage of complementarities between demand for a market commodity and a related non-market environmental resource. For example, the demand for fishing is probably a function of water quality since the latter affects fish populations. An improvement in water quality may cause the demand for bait to increase from D^0 to D^+, while a reduction in water quality may cause demand to decrease from D^0 to D^- (Figure 3.3). Data on exactly how the price and quantity of bait change as water quality changes could be used to estimate the demand for water quality.

Another market-related technique takes advantage of situations when someone actually buys a combination of commodities in a single package. For example, consider your decisions when buying or renting a house. You probably evaluate the importance of the number of bedrooms, the size of the lot, and its distance to where you work. In addition, your willingness-to-pay may be based on the environmental resources that the property's location provides. These include distance to a public beach, whether there is a water view from the property, the amount of water frontage, the amount of privacy, and so on. The *hedonic price technique*

Price

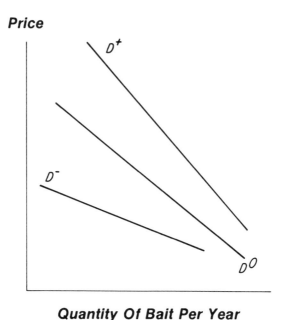

Quantity Of Bait Per Year

Figure 3.3 Effect of water quality on demand for bait: D^0 is original demand; D^- is demand with reduced water quality; and D^+ is demand with improved water quality.

is designed to determine the economic value of housing attributes, including environmental resources (Figure 3.4). The technique estimates the implicit or shadow value of environmental resources capitalized in property values and the demand for these resources. It has been used to study the value of air quality, water quality, noise reduction, visual amenities, and proximity to recreational sites. In addition, it could be used to predict changes in property values when characteristics of the environment change.

A third market-related method, the *travel cost technique,* is used extensively to study the demand for outdoor recreation. It uses information on travel expenses, number of visits to a particular recreational site, and other economic data on income, tastes, and prices of substitutes. Travel expenses are a proxy for the "price" of a visit; the farther you live from the site, the greater the travel cost and, therefore, shadow value (Figure 3.5). There are many potential applications of this technique to assess recreational values within the coastal zone, as well as opportunities to

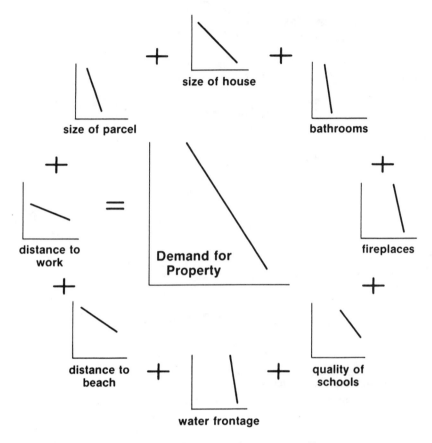

Figure 3.4 Hedonic demands for property attributes.

assess the impacts of externalities. For example, the technique could be
used to estimate the demand for ocean beaches and the effect of substitutes, traffic, and degree of crowding on demand (Figure 3.6).

It should be mentioned that the value of time is an important consideration in travel cost studies. That is, time spent traveling and at a particular recreational site has an opportunity cost since you "cannot do two
things at once." Thus, the income equivalent of time-costs must be considered as well as out-of-pocket expenditures such as on gasoline.

Each of the above market-related approaches has relative strengths and
weaknesses which must be weighed when selecting the appropriate method
for a particular job. It is also very important to try and include the influ-

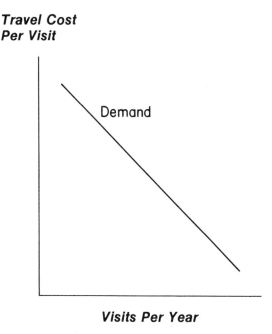

Figure 3.5 Demand based on travel costs.

ence of substitutes, congestion, and income on demand since they will influence estimates of surplus. See Anderson and Bishop (1986), Kneese (1985), Freeman (1979), and Randall (1981) for more details on these methods, as well as Chapters 5, 7, and 8 in this book.

Non-Market Methods

Sometimes there are insufficient market and market-related data to use the above techniques. For example, the travel cost technique is undermined when distances to the recreational site and, therefore, travel costs are negligible. In other cases, people may not perceive certain externalities and non-exclusive benefits, let alone changes in their levels. This is often true of water quality since changes in chemical and microscopic properties of water cannot be distinguished directly with the naked eye. In these situations, valuations are not revealed by behavior. In addition, many effects are not expected until the future and, therefore, are not reflected in current behavior. Finally, certain classes of economic value,

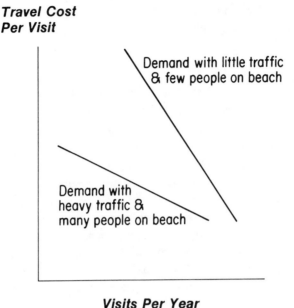

**Travel Cost
Per Visit**

Demand with little traffic
& few people on beach

Demand with
heavy traffic &
many people on beach

Visits Per Year

Figure 3.6 Effect of crowding on beach demand.

such as the non-use values, are not revealed in behavior by definition. In these situations, economists usually use various forms of contingent choice methods, particularly the *contingent valuation method.*

The contingent valuation method is a survey technique that is designed to collect economic data directly from individuals. A hypothetical, yet realistic, market for an environmental resource is described to respondents. Typically, respondents are asked for their maximum willingness-to-pay to prevent a decline in environmental resources or to support an improvement. The method can be used to estimate directly either total beneifts or only surplus. The valuations are contingent on the conditions described in the hypothetical market—hence, contingent valuation. There are several potential sources of problems awaiting the researcher, but most can be handled.

The contingent valuation method is often used to assess the economic value associated with large changes in the levels of water quality, visibility, wildlife encounters, noise, and other environmental resources. It is also a necessary approach for the assessment of existence values and is very useful for assessing the value of potential future events before they occur. See Anderson and Bishop (1986) and Chapters 8 and 9 in this book for more details.

Comparative Methods

This section discusses a few methods used by economists to organize economic information on benefits and costs of public policies. *Benefit-cost analysis* and *cost-effectiveness analysis* are concerned with economic efficiency. The former compares alternatives on the basis of net economic benefits. Thus, a project will fail or pass the benefit-cost criterion depending on whether efficiency in the allocation of resources is expected to be reduced or increased. In addition, benefit-cost analysis can determine the efficient size of a public project or program.

In cost-effectiveness analysis, one searches for only the least expensive means to achieve a stated goal; whether benefits exceed costs is not a consideration here. For example, society might decide that the potability of groundwater must be protected or that beaches must be made available to the public. The economists' task in using cost-effectiveness analysis is to determine which alternative has the lowest present value of costs. Using the groundwater example, is it cost-effective to regulate land uses that affect groundwater quality, to build tertiary treatment plants, to treat water at the well, to use the assimilative capacity of wetlands, to use spray irrigation, to require households to install denitrification systems, or to buy bottled water?

The third method considered here—*economic impact analysis*—is a detailed accounting of the economic benefits and costs of policy options. This approach disaggregates the benefits and costs of an action across affected parties. Information from an economic impact analysis allows decision-makers to consider the redistribution of income and wealth which is expected to result from a project. As we shall see, transfers of economic surplus from one group to another have no implications for the results of efficiency analyses; however, they are very important considerations for policy makers who must consider equity as well as efficiency.

This stand on equity might sound callous. In fact, many individuals criticize the use of benefit-cost analysis because it does not consider the implications of the distributions of benefits and costs, whether they be across individuals in the current population, across generations, or involve effects on wildlife. However, much of the criticism is actually misplaced. Economists do not pretend to judge equity or what is right or wrong. But the omission of such judgments from economic analyses should not suggest that economists do not recognize their potential importance. This can be understood if we make distinctions between *positive and normative analyses*. Positive analysis is concerned only with the facts. What *is* the demand for public beaches? What *is* the change in surplus if public access to beaches is doubled? What *are* the opportunity costs of increasing public access? Contrast these with normative analyses which attempt to prescribe what *should* be done based on certain ethical principles of fairness and rights. Thus, "we *should* acquire more beaches for public use" and "we *should* protect the potability of groundwater at any cost" are normative statements.

Economists struggle to concentrate on the positive side of policy analysis. As a profession, we can speak authoritatively on the implications of resource allocation for efficiency, but can say no more about what should be done than the next person. I quickly add that no scientist—be he an ecologist, chemist, physicist, or political scientist—can preach what should be done based on their studies of environmental and social systems. These are largely matters of philosophy and personal opinion.

I do not mean to suggest that economic analyses do not have ethical implications. You might have noticed that willingness-to-pay is constrained by one's ability to pay—in other words, income. From a pragmatic point of view, however, our willingness-to-pay to promote nonmarket values will be determined by our incomes and not by what we or others might want them to be.

In what follows, I shall refer almost exclusively to benefit-cost analysis since the concepts and practices are the same for cost-effectiveness analysis. Comments on economic impact assessments will be made throughout.

Benefit-cost Analysis

Benefit-cost analysis systematically identifies and organizes economic benefits and costs that are expected to result from a proposed public policy or program, including the implicit policy alternative of doing nothing (i.e., the status quo). As indicated above, much confusion surrounds benefit-cost analysis in practice versus benefit-cost analysis in principle. There is no doubt that benefit-cost analysis has been misused and abused. In many cases, it is easy for someone to come up with the answer that he wants by "fudging" with discount rates or by selectively omitting certain opportunity costs and damages, particularly those associated with non-exclusive resources and externalities. However, this is a problem with certain practitioners and not an inherent flaw of the method. I ask you to reconsider the merits of benefit-cost analysis based on its proper usage.

Benefit-cost analysis is based on what economists call the *potential Pareto criterion*. Early in the nineteenth century, Italian economist Vilfredo Pareto argued that a public policy which changes the allocation of productive resources is beneficial if no one is hurt by the policy and at least one person gains from it. In the real world, however, this result is rare. As an alternative, economists Sir John Hicks and Nicholas Kaldor argued for the relevance of potential compensation in public finance—a reallocation of resources increases efficiency when the economic benefits exceed the opportunity costs regardless of who incurs the benefits and costs. In these cases, winners could *potentially* compensate losers—hence, the potential Pareto criterion.

Notice that I did not claim that winners *should* compensate losers. This may or may not be true depending on the law and what society believes is just or fair. Again, goals concerning equity and fairness are extra-economic goals. Thus, benefit-cost analysis does not provide automatic decisions, nor does it supplant the political decision making process.

While reading this section, consider an ambitious benefit-cost analysis of the development of a coastal region under current land use zoning. Let the baseline situation to be present and expected future development under current zoning. The overall question is, What are the net economic benefits of managing economic growth in the coastal zone? In order to answer this question, the net economic benefits of expected growth must be compared to the net economic benefits of alternative uses of coastal resources.

You can imagine the arguments, pro and con. The construction industry and local businesses might argue that development will generate considerable income and employment in the coastal zone. Furthermore, part of the income will be respent, thereby having a multiplier effect on the coastal economy. In addition, the "improvement" of the vacant land will increase the tax base of the towns and, therefore, increase tax revenue.

On the other hand, many residents might argue that population growth increases town expenses, causing tax rates to increase. Also, the sewage from additional households will pollute groundwater and nearby estuaries and lagoons. Traffic will worsen considerably. Beaches will become overcrowded. Public access to open space will be reduced significantly. Critical wildlife habitat will be lost. The environmental services that wetlands provide to communities will be chipped away. Sound familiar? It does if you live in or near the coast.

The remainder of this chapter highlights concepts and practices related to these issues. Chapter 10 illustrates this discussion with a hypothetical benefit-cost analysis of development on Cape Cod, Massachusetts.

Social Discounting

The previous discussion of discounting concerned private, or personal, discount rates, based on goals of utility or profit maximization. Many economists argue, however, that private rates are higher than those for discounting social benefits and costs. Unfortunately, this literature can be very complex and sometimes arcane and, therefore, will not be covered here. Suffice it to say that many economists argue for real (i.e., non-inflationary) social discount rates of between 2% and 6%.

Benefit-cost analysis uses the present value criterion to organize economic benefits and costs that occur across groups and across time. For example, the data in Table 3.1 for the economic effects of three alternative policies for protecting shoreline would be difficult to interpret in their present form even though the time period is quite short. However, we can discount all future economic benefits and costs to their present value using the social discount rate. The *net present value* (NPV) is then defined as the difference between the present value of economic benefits minus the present value of economic costs. Mathematically, this is equivalent to the following operation:

$$NPV = \frac{(B_0 - C_0)}{(1 + d)^0} + \frac{(B_1 - C_1)}{(1 + d)^1} + \frac{(B_2 - C_2)}{(1 + d)^2} + \frac{(B_3 - C_3)}{(1 + d)^3}, \quad \text{or}$$

$$= \sum_{t=0}^{t=3} (B_t - C_t)/(1 + d)^t$$

where t is the year in which the benefits (B) and costs (C) occur, d is

Table 3.1

Net Present Values (NPV) of Three Alternative Investments for Shoreline Protection. Values are hypothetical. The social discount rate is 4%.

Alternative	Category	Time Period			
		t = 0	t = 1	t = 2	t = 3
I	Benefit, B_I	15	40	50	50
	Cost, C_I	80	40	10	10
	$B_I - C_I$	−65	0	40	40
	NPV_I		— $7.6 —		
II	Benefit, B_{II}	20	40	40	40
	Cost, C_{II}	30	30	30	30
	$B_{II} - C_{II}$	−10	10	10	10
	NPV_{II}		— $17.7 —		
III	Benefit, B_{III}	5	50	50	50
	Cost, C_{III}	60	60	5	0
	$B_{III} - C_{III}$	−55	−10	45	50
	NPV_{III}		— $21.4 —		

the social discount rate, and Σ is a summation sign indicating the addition across years. The net present values in this example using a 4% rate of social discounting (i.e., d = 0.04) are $7.6 million for alternative I, $17.7 million for alternative II, and $21.4 million for alternative III. Based on this analysis, alternative III is the most efficient alternative for shoreline protection since its net present value is the greatest.

In general, net present value can be calculated from

$$NPV = \sum_{t=0}^{t=T} (B_t - C_t)/(1 + d)^t$$

where T is the last year to be considered. Few projects include analyses beyond 50 years since present values decrease rapidly with time (recall Table 2.2). Therefore, the timing of benefits and costs are crucial to the determination of net present values.

Sunk Costs, Transfer Payments, and Double Counting

It is easy to include irrelevant costs and to count benefits and costs more than once even when you are careful. Of course, this practice must be avoided; otherwise, the net present values of a project's alternatives (or the present values of costs in cost-effectiveness analysis) would be biased, leading one to erroneous conclusions.

Certain "costs" are not opportunity costs. A social opportunity cost is assigned to a project at the time it is incurred and only if it involves the use of productive resources. Thus, costs which might be related to an overall policy but which were incurred in the past are *sunk costs*. Since they occurred in the past and have nothing to do with present or future costs (except perhaps indirectly if they lower the available budget or otherwise limit the policy alternatives that you can choose from), they should be ignored.

Depreciation costs are also irrelevant to economists. They are simply accounting devices for tax purposes.

On the other hand, *transfer payments* may involve actual economic benefits and costs. Nevertheless, they have no influence on a project's efficiency or cost-effectiveness. As the term implies, a project's alternative might involve the transfer of income and benefits from one group to another but with no change in net benefits—a *zero sum* result. Taxes and subsidies usually fall into this category. For example, the land transfer tax that some communities charge on new construction is a transfer of surplus from consumers and producers of new houses to the communities as a whole. (To be precise, there is actually an associated "deadweight" loss, although it is usually relatively small. We will consider this loss category in Chapter 4.) Similarly, increases in property taxes necessitated by population growth are simply transfers from current property owners to the overall population. Of course, such transfers create winners and losers, and although it makes no difference on efficiency grounds (recall the potential Pareto criterion), the effect may be contrary to political or ethical beliefs about equity. Again, an economic impact assessment provides information to decision makers which goes beyond the needs of benefit-cost analysis and cost-effectiveness analysis.

Double counting occurs when economic benefits or costs are counted more than once. It would be double counting if a policy that reduces costs is counted once on the cost side by lowering costs and again as an increase on the benefit side in the same calculation. In situations like these, it helps to distinguish between costs and damages (i.e., reduced benefits). Probably more common, however, is to count benefits or costs once as a flow that occurs over time and then as a stock. An example will clarify the point. Suppose a highway expansion is being considered in order to reduce travel time to a state beach. As part of the benefits calculation, the state includes the value of time saved over the next 50 years for motorists due to reduced travel times. It would be double counting, however, to include increases in property values that result directly from the improved roads.

The treatment of transfer payments can result in double counting too. The possibility depends largely on what you consider to be the boundary of the analysis (boundaries will be discussed below as part of external and internal effects). For example, the United States has subsidized insurance for waterfront properties. Although waterfront property owners undoubtedly consider the amount to be a benefit, from the country's point of view, the disbursement is a transfer of tax dollars from all Americans to a few wealthy owners.

Direct and Indirect Effects and Related Concepts

First consider the difference between *direct,* or *primary,* and *indirect,* or *secondary, effects.* Direct effects concern the intended output of a project. Using our example, construction companies add to the supply of housing—housing is the output. Meanwhile, the demand for housing continues to grow rapidly. The direct net benefits are the present value of maximum total willingness-to-pay by households for the houses, minus the present value of the opportunity costs of the resources used to develop vacant land–labor, materials, construction equipment *and* the earnings that owner(s) of the construction companies forego by not building or investing their capital elsewhere. As we saw in Chapter 2, this difference is equivalent to consumer surplus plus economic profit.

Any other expected effects induced by, or stemming from, the proposed subdivision are indirect effects. The question is whether indirect effects represent real additions to or deletions from the value of society's output. In other words, do the indirect effects represent gains or losses in consumer surplus or economic profit, or are they simply transfers of benefits from one group to another?

Consider the earnings of resources used in construction, particularly of labor but also of businesses that sell building materials. *Unless the development employs labor and resources that otherwise would be unemployed, the net benefits of their use induced by construction is zero.* Why? Because the resources would still be employed elsewhere in the production of other commodities valued by society—once again we encounter opportunity costs.

The same conclusion holds for claims of respending induced by and stemming from construction. Some argue that those employed by the construction industry and others selling materials and services (e.g., real estate agencies, insurance companies) to the industry will respend part of their earnings in other nearby markets—restaurants, grocery stores, clothing stores, and so on. Again, the earnings induced by construction represent net economic gains only if the resources used in the production of these inputs would otherwise be unemployed. The *net* indirect effects drop off rapidly after the first round of spending in economies which are near full employment.

This observation about opportunity costs is important to understand since much of the alleged benefits of a project are in terms of respending. *Multipliers* of between three and five are usually misused. Although the earnings of factors employed by a project may be spent and respent three to five times as the income moves through the coastal economies, it is necessary to subtract the opportunity costs at each step to determine the net effects. Suffice it to say that the multiplier effect will usually be considerably less than one when unemployment is low.

One must be careful when treating the opportunity costs of labor in benefit-cost analysis and cost-effectiveness analysis. Even if there is considerable unemployment, it is necessary to determine whether the project(s) will actually hire the unemployed. Furthermore, it is necessary to consider whether present unemployment levels will persist throughout the length of the project.

The same can be said of other earnings stemming from the project. Businesses such as new restaurants, new grocery stores, new clothing stores, and new real estate agencies might be built to service the new residents. Again, one must inquire about the opportunity costs of these ventures.

Externalities can be a form of indirect effects too. One form concerns municipal finances. Clearly, the increased market value of the developed land will raise the tax base of towns, generating additional tax revenue unless tax rates are reduced commensurately. However, it is not clear that this represents a net benefit. Obviously, towns are required to provide certain services to new property owners. These might include police and fire protection, road maintenance, trash removal, and public education. Furthermore, new types of services such as investments into elderly services and welfare programs may be required as towns grow. The additional costs of providing these services may exceed the additional tax revenues generated by development. For example, the capital expenses of new schools and new police cars are quite high; hence, everyone's tax rate increases.

Of course, from a pure efficiency standpoint, it does not matter who bears the costs of increasing town services. Thus, the *increased* taxes paid by current residents is, in part, a transfer to the new residents—a pecuniary externality. Nevertheless, it might matter very much to those paying the higher taxes. Again, economic impact assessments provide additional, useful information to decision makers, even though transfers are superfluous to efficiency analysis.

Other pecuniary externalities are worth mentioning. For example, the "discovery" of the coastal zone by residents of nearby urban areas has increased the market demand for coastal properties dramatically during the past decade, causing prices to increase sharply. In addition, the increased property values of existing houses stress the ability of low- and middle-income families and those on fixed incomes to pay their property taxes. Finally, the higher prices make it increasingly difficult for young couples to purchase or rent their first house. Economic impact assessments can quantify these impacts.

However, many of the externalities of development are technical externalities which affect utility and production directly through changes in the quantity and quality of coastal resources. In many cases, they cause real losses in personal utility and productivity and, therefore, must be included in a benefit-cost analysis. Until recently, most of these effects were *incommensurable*—that is, beyond measurement. However, the market-related and non-market methods such as those introduced earlier in this chapter can be used now to quantify these damages.

An example of a technical externality that affects utility directly is reduced groundwater quality (see Chapter 9). Losses in satisfaction due to diminished recreational opportunities are also indirect damages of development (see Chapters 7, 8, and 10). This may be due to reduced water quality in estuaries, reduced access to recreational sites, or increased crowding at remaining sites. Additional time spent in traffic will lower most peoples' well-being and possibly the profits of some businesses since potential customers might be reluctant to cross or merge onto crowded streets. Finally, traffic can reduce the productivity of firms when transportation time for businesses increases.

The "Art" of Benefit-cost Analysis

Despite the rigorous framework, considerable judgment usually enters into applied benefit-cost analysis. However, economists can strengthen the analysis and make it more useful to decision makers by providing a range of likely estimates of net present values and by consulting with experts who can provide professional advice on technical matters which influence the estimation of benefits and costs. For example, oceanographers can advise economists on the timing and probability of changes in relative sea level (see Chapter 6) and of a salt water intrusion into public wells. Similarly, dredging operations may affect the growth and quality of shellfish stocks. This kind of information is important since future costs and damages are discounted to present values.

The different types of judgments essentially follow the steps in a benefit-cost analysis. first, the analyst must decide on the boundary of the analysis. That is, which benefits and costs will be *internal* to the project and which will be *external,* or excluded. Often this is determined by the governmental body or person who pays for the study. There is no reason to believe, however, that all groups affected by a policy fall within convenient geopolitical boundaries. It is the economist's job to point out which categories of social costs and benefits are being excluded and, possibly, which internal benefits and costs are actually transfers.

The second step of benefit-cost analysis involves identifying relevant alternative policies and the specific system of accounts. The unit that commissions the study probably will determine the specific alternatives. However, the analyst probably will decide on the specific categories of economic benefits and costs, perhaps constrained by what is considered to be internal to the study. Decisions on admissible indirect effects and the incidence of transfers and double counting are not always obvious.

The next step is to collect data on the benefit and cost components. This is always easier said than done since market prices do not always reflect opportunity costs or utility at the margin. Furthermore, many benefits and costs are not captured by markets. The economists' professional judgment always comes into play here.

Related to this step, time and budget constraints on the analysis often force approximations, concentrating only on the expected changes in the flows of income with a project vis-a-vis without a project. Among other things, one often makes the tacit assumptions that (1) markets are in competitive equilibrium; therefore, prices reflect social values of commodities and inputs at the margin; (2) changes in prices and/or quantities are so small relative to the entire market that they can be valued as marginal changes; and (3) the marginal utility of a dollar is constant regardless of income levels and is equal across individuals. Clearly, these assumptions will be violated in many cases. For example, the many non-market effects discussed above would not be included in such an assessment. When important direct effects are being overlooked, the decision maker might judge whether the net present value of a project causing these effects would be outweighed by the unaccounted damages. This approach is a poor substitute for actual quantification, however.

Next, an economist tries to anticipate future events in order to determine the magnitudes of benefits and costs through time—all without the aid of a crystal ball. Ideally, we want to know what conditions would exist *ex ante* as a result of a policy alternative and compare these to conditions that would exist without the project. How will peoples' tastes and preferences change? How will firms react to the policy alternative? How will population size change? What will the distribution of income be through time? What will the unemployment rates be? Inflation rates? Market structures? And so on. Clearly one's confidence in the assignment of benefits and costs will lessen with distance into the future. In practice, analysts sometimes resort to describing conditions before and after a project. Thus, present conditions are extended into the future to represent costs and benefits without the project, and expected changes in present conditions due to the project are projected into the future to represent with-project benefits and costs.

The final step involves *sensitivity analyses* to learn how estimates of net present values change when the social discount rate changes and when relatively large benefit and/or cost components vary in magnitude and timing. As indicated above, economists do not know *the* social discount rate. Therefore, it is important to demonstrate the sensitivity of results to different values for the discount rate. Furthermore, estimates of benefits and costs will be imperfect and, therefore, be associated with some degree of error or uncertainty. Again, an economist should undertake sensitivity analyses of how estimates of net present values change as estimates of economic benefits and costs change. It would be a mistake, however, to conclude that discount rates and benefit and cost estimates are determined arbitrarily. As indicated above, the social discount rate is thought to be between 2% and 6%. The use of rates beyond this range might be inappropriate for the analysis of public investments. Also, an experienced economist will have an educated idea of the likely ranges of values for some benefits and costs.

Chapter 4

"Thinking Economically"

"Thinking economically" (a phrase coined by M. Levi (1985) in his book) about the allocation of coastal resources is the purpose of this chapter. The following parables inculcate the concepts in previous chapters and reveal various ways in which economics can be used to analyze resource conflicts. this is accomplished by evaluating 10 stereotypic arguments that you might hear or read about in newspapers. In addition, a few new concepts are introduced in order to increase your knowledge of economics. To help you through this chapter, you are directed to the index to locate where the economic concepts were first described.

1. A large construction company has applied for a building permit to construct a 200-unit condominium complex in Boomtown. The designated area is adjacent to a quarter-mile of ocean beach, near a coastal lagoon that is used for shellfishing and is within the recharge area of the town's public well. At a town meeting, the company's president explained that the project will employ 25 skilled workers from Boomtown during the three-year construction phase and 15 people thereafter to manage and run the complex. He emphasized that the construction phase of the project will pour $2.4 million into the local economy over the next three years, including wages for the 25 workers. Furthermore, he claimed that the money will be respent several times locally on building materials, food, housing, and at sundry other businesses. Using a multiplier of 3.5, the construction phase is expected to generate $8.4 million in local income. What a deal!

Alarms should be going off in your head. Indirect effects. Opportunity costs. Discounting. Large construction companies tend to argue from the input side or according to the project's expenditure on factors of production (particularly labor), including multiplier effects. I am not suggesting that this is inadmissible—just woefully incomplete.

First, this argument does not even consider the net direct benefits of the condominium project. Certainly, the owners of the construction company expect to cover their opportunity costs (i.e., the net income from investing their capital in construction elsewhere) and presumably make an economic profit. And what about the expected value of consumer surplus for those who will buy the condominiums? Both omissions clearly ignore the net direct benefits of the project.

On the other hand, the concentration on expenditures induced by the project in markets for labor, building materials, and housing most certainly overestimates the *net* indirect benefits induced by the project. The opportunity costs of inputs used in the production of the condominiums as well as in local markets where employees respend their income must be deducted to determine net indirect benefits. In one extreme case when all inputs would be employed with or without the project, the indirect benefits would be near zero. This is most likely to be the case when skilled labor is employed. If the net multiplier for this example is 10%, the net indirect benefits would be at most $2.4 million $\times$ 0.10, or, $240,000. However, even this figure would be an overestimate because we did not discount the net benefits in the second and third years to their net present value. If the net indirect benefits are spread evenly over the three years, the net present value is $228,800 using a 5% social discounting rate. (Note that for simplicity, I treated the net benefits for each year as if they were received on the first day of the year, hence, the years are 0, 1, and 2 instead of 1, 2, and 3. In reality, the benefits would be spread evenly across the three years on almost a daily basis. Although a different formula would be used, the discounting principle is the same.)

Finally, there is no accounting of the technical externalities which the development might create. Would traffic in Boomtown be increased significantly? Would groundwater within the recharge area remain potable? Would shellfish beds in the nearby lagoon be overfished or closed due to sewage pollution? Would the public lose access to the stretch of beach adjacent to the planned complex? These, too, are possible external impacts of the construction project that need to be included in a benefit-cost analysis.

2. An entrepreneur wants a permit to dredge a 100-slip marina in Monopoly Cove. The owner of the only marina now in the cove argues,

however, that this could put him out of business. Without being specific, his accountant estimates losses of over $100,000 a year. In addition, two employees might be laid off.

Firms that expect to be affected by increased competition tend to argue from the output side. That is, they announce the amount of revenue (not economic profit) that they expect to lose. There is no doubt that new entry into a market could reduce profits. Although the "fairness" of increased competition is left to others to judge, it is possible to estimate the economic benefits and impacts of the proposed entry.

In this case, the $100,000 in expected losses in revenues is from the rental of slips and the sale of supplies to the present firm's clientele. It is not a net annual loss in economic profit. From this we must deduct the opportunity costs of inputs that the firm uses to provide services to customers and the opportunity cost of his land. For example, vacant slips might have value as a fishing pier or a boat launch. Even if the slips are not rented for a few years, one must establish when they might be filled in the future as the demand for slips increases with a growing coastal population.

Next, consider the opportunity costs for the two people who might be laid off. Would they be employed elsewhere such as the proposed marina? Their opportunity costs in the labor market must be deducted from their present earnings in order to estimate a net loss.

The possibility that the owner might go out of business raises several interesting issues. It is important to consider the difference between accounting profit and economic profit. The accountant's figures might show that the owner earns $60,000 annually. But what part of that represents his opportunity costs, including perhaps the marina itself if he owns the land and capital? If he would actually go out of business, it might suggest that he could still earn $60,000 a year by working elsewhere, by leasing the marina to someone else, and/or by selling it to another business. Of course, if he is making an economic profit, that would be considered a social loss. There is also the possibility that the owner derives personal utility directly from time spent working at the marina. The loss of this satisfaction should be considered. Finally, one should consider the owner's fixed costs and transaction costs if indeed he goes out of business, including closing fees, lawyer fees, and possible moving costs.

Finally, consider the possibility that the owner does not go out of business and that the loss in revenue is due solely to a lower price or rental fee for slips. Also, for the sake of argument, let's say that the existing and proposed marinas are identical and that the expected lower rental fee is just enough to cover all variable and fixed costs, including the respective owners' opportunity costs. Thus, without the second marina, the existing firm is making an economic profit that it will lose with the second firm's entry. However, the lower price also means that the current customers will benefit with a gain in consumer surplus. Furthermore, the new marina and lower price will allow more people to dock their boats, resulting in an additional increase in consumer surplus. Thus, the loss in the existing firm's profit is simply a transfer from the producer to consumers—a pecuniary externality. In addition, though, there is a further gain in economic surplus composed of an increase in consumer surplus that exceeds the loss in economic profit. Ah, the joys of perfect competition. These are direct effects since they occur in the market that would be affected directly by entry.

One way to illustrate these qualitative results is with a *monopolistic firm*. Unlike pure competition, monopoly simply refers to a market structure for a commodity where there is a single producer who sets market prices. Although the economists' model of monopolistic behavior still assumes profit maximization, marginal revenue is not constant and equal to price as it is for a firm in a competitive industry. To see this, consider the following equation for the local market demand for boat slips:

$$\text{Slips} = 115 - (0.05 \times \text{Price})$$

Quantity demanded is one when the annual rental price is \$2,280 (115 − 0.05 × \$2,280 = 1); total revenue is also \$2,280 in this case (Table 4.1). At \$2,260, quantity demanded increases to two (115 − 0.05 × \$2,260 = 2) and total revenue is \$4,520 (2 × \$2,260 = \$4,520). Marginal revenue, or the gain in total revenue from renting the second slip, is \$2,240 (i.e., \$4,520 − \$2,280 = \$2,240). At \$2,240 per slip, quantity demanded is three, total revenue is \$6,720, and marginal revenue is \$2,200 (i.e., \$6,720 − \$4,520). The process continues. When the rental price is \$1,480, quantity demanded is 41 and marginal revenue is \$680.

Let the marina's average cost for variable inputs be:

$$\text{Average variable cost} = 496 + (2 \times \text{Slips})$$

For example, the average variable cost of providing the first slip is \$498

Table 4.1

Marginal Revenue and Marginal Cost Account for a Profit-Maximizing, Monopolistic Firm (FC = fixed costs)

Price	Number of slips	Total revenue (TR)	Marginal revenue (MR)	Marginal cost (MC)	Average variable cost (AVC)	Total variable cost (TVC)	Profit = TR − TVC − FC	Marginal profit (MR-MC)
$2,280	1	$ 2,280	$2,280	$498	$498	$ 498	−$1,580	$1,782
2,260	2	4,520	2,240	502	500	1,000	158	1,738
2,240	3	6,720	2,200	506	502	1,506	1,852	1,694
2,220	4	8,880	2,160	510	504	2,016	3,502	1,650
2,200	5	11,000	2,120	514	506	2,530	5,108	1,606
2,180	6	13,080	2,080	518	508	3,048	6,670	1,562
2,160	7	15,120	2,040	522	510	3,570	8,188	1,518
2,140	8	17,120	2,000	526	512	4,096	9,662	1,474
2,120	9	19,080	1,960	530	514	4,626	11,092	1,430
2,100	10	21,000	1,920	534	516	5,160	12,478	1,386
	—							
	—							
	—							
1,500	40	60,000	720	654	576	23,040	33,598	66
1,480	41	60,680	680	658	578	23,698	33,620	22
1,460	42	61,320	640	662	580	24,360	33,598	−22

a year (496 + 2 × 1 = $498). The average variable cost for the second slip is $500 (496 + 2 × 2 = $500), while total variable cost is $1,000 (2 × $500). Marginal cost, or the increase in the total cost of providing the second slip, is $502 ($1,000 − $498). As above, the process continues. The marginal cost of the forty-first slip is $658 (Table 4.1).

(The reader who is familiar with differential calculus knows that the values for marginal revenue and marginal cost will differ slightly if you derive marginal revenue and marginal cost curves from the above equations. The slight difference is due to evaluating these values at a point [e.g., *at* 41 slips] as opposed to subtracting values between two points as I am doing here. Using differential calculus, marginal revenue and marginal cost are exactly equal at 41 slips.)

What is so special about the provision of 41 slips? As can be seen in Table 4.1, the firm maximizes profits at this level of supply. That is, given the nature of the demand curve (i.e., what the consumers are willing to pay) the additional revenue (i.e., marginal revenue) from providing the forty-second slip ($640) would be less than the additional cost (i.e., marginal cost) of providing it ($662). Therefore, total profit would decrease if more than 41 slips are supplied.

Using this stepwise method, consumer surplus is $16,400 at 41 ships. Economic profit is total revenue ($60,680) minus the total cost of variable ($23,698) and fixed factors of production. If *fixed costs* are $3,362, the monopolist's profits are $33,620, and total net economic benefits are $50,020 (the sum of consumer surplus and the monopolist's profits).

Notice that the monopolist's marginal cost curve is *not* a supply curve. That is, the marginal cost curve does not delineate quantity supplied at each price as it does for firms in a competitive industry. At $1,480 per slip, the monopolist supplies 41 slips while a firm in a competitive industry would supply 246 slips. (Recall that firms in a competitive industry supply commodities until their marginal cost of production equals the market price of the output. In this case, price = marginal cost = $1,480. The reader is invited to extend the results in Table 4.1 to verify this result.)

There is another reason to emphasize the difference between marginal cost ($680) and market price ($1,480) under monopolistic conditions. Under non-competitive market structures, the marginal cost of inputs is less than society's valuation of the commodity. (See any text in microeconomics such as Mansfield's [1979] for a discussion of other market structures.) Thus, data on production costs would underestimate the marginal value of output to society as revealed by market price. This difference is in addition to efficiencies due to possible technical externalities.

Next, suppose that entry by the proposed marina would establish a competitive market structure. In this case, the market supply is in fact the summation of the two (identical) marginal cost curves as drawn in Figure 4.1c:

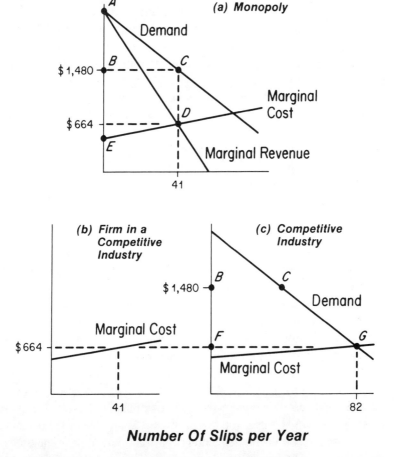

Figure 4.1 Markets for marinas under (a) monopolistic and (b) competitive conditions.

$$\text{Slips} = -248 + (0.5 \times \text{Industry marginal cost})$$

Each marina would take the price determined by the market; the first marina could not maintain higher prices because its customers would simply go to the new marina. The $660 market price is determined by equating the demand and supply equations and solving for price, recognizing that price and "industry marginal cost" variables are the same when the market is in equilibrium. At this price, demand and supply are in equilibrium at 82 slips, with each marina supplying 41. (The price $660 is slightly more than the marginal cost of a firm at 41 slips due to rounding errors.) The fact that economic profit is zero for each firm (total revenue is $27,060, total variable cost is $23,698, and fixed cost is $3,362) indicates that each firm is just covering its opportunity costs. And although the original firm would lose an economic profit of $33,620, the combination of a lower market price and increased output increases consumer surplus by $50,020, from $16,400 to $66,420. Since the increase in consumer surplus exceeds the loss in the first marina owner's economic profits, entry into this market represents a potential Pareto improvement for society as a whole.

This example was designed to drive economic profits to zero with entry of a single firm. However, if the two firms still made economic profits, one might expect more entry depending on possible barriers to entry and the opportunity cost of shoreline property. Entry would lower price further until each firm is just covering its opportunity costs. In fact, we could have used such a scenario to characterize the qualitative results discussed above. For a further discussion of adjustments in competitive markets, see Mansfield (1979) or another intermediate text in microeconomics.

3. One possibility for raising money to acquire open space and protect against certain external effects of development is a transfer tax—a tax on the sale of newly developed properties. Some people argue that a tax will have a negative impact on the housing market. Let's think economically about this issue.

As we shall see, there will be a loss in surplus, at least in the short run. However, a better question to ask is, what is the size of the potential impact and how is it distributed among construction companies and buyers?

Consider a 2% tax on the sale of new houses. Let's say that at a market price of $95,000, 1000 new houses would be sold this year in your coastal zone (Figure 4.2a), generating $95 million in total revenue for construction companies. Recognize that the commission for real estate agencies is part of the construction companies' marginal costs.

A 2% transfer tax lowers the *effective* demand for new houses (Figure 4.2b). Although peoples' willingness-to-pay remains the same, part of their payment will be in the form of the tax; hence, the demand curve faced by the construction companies actually moves down. In this example, quantity demanded declines about 7.6% from 1000 to about 924 houses. In addition, market price declines to about $94,620, although the total price to consumers (including taxes) increases by 1.6% to $96,520. Total expenditures by consumers decline to about $89.2 million (924 × $96,520). Total revenue for construction companies declines by $7.6 million to about $87.4 million (924 × $94,620). The tax program receives $1.7 million [924 × (0.02 × $94,620)] for the acquisition fund.

However, in order to estimate the annual economic impact on construction companies we need to look at the change in economic profit caused by the tax. In this hypothetical example, total variable costs from the production of 1,000 houses is about $92.5 million (this is the area under the industry's marginal cost curve between 0 and 1,000 houses). To this, add $1.5 million in fixed costs that the firms incur regardless of their level of construction. Therefore, economic profit is about $1 million initially ($95 million in total revenue minus $94 million in total costs). In contrast, the total variable cost of building 924 houses is about $85.3 million. Since fixed costs stay the same by definition, regardless of levels of output, profit is reduced to about $0.6 million ($87.4 million − $85.3 million − $1.5 million) for a total loss for all firms of $400,000. (You might have noticed that we did not need to know fixed costs to estimate *changes* in benefits since fixed costs are constant by definition. However, if any firms go out of business, their fixed costs are added as an additional loss.)

Figure 4.2b tells us more, however. Area ABD, or $73,000, is the only true loss to society as a whole. In economics, this loss due to taxes is called a *deadweight loss*. Part of it is attributable to a loss in profits (area BCD, or about $15,000). The other portion is a net loss in consumer surplus (area ABC, or about $58,000).

The rest of the reduction in profits ($400,000 − $15,000 = $385,000) and in consumer surplus [($96,520 − $95,000) × 924 = $1,404,480] is actually a transfer in the form of a tax from this market to the government

Price

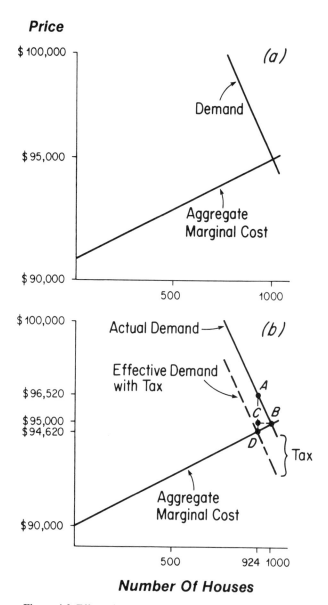

Figure 4.2 Effect of transfer tax on market demand for houses.

and, therefore, society as a whole. It is not a loss to society, although the buyers and sellers in this market are clearly affected negatively.

So far we have been engaged in a *partial equilibrium analysis*. That is, we have considered only the immediate effects of the tax in the market for houses and not any factors that might feed back and change market conditions. A consideration of possible adjustments in related markets for inputs, substitutes, and complements and subsequent feedbacks is called *general equilibrium analysis*. Rather than illustrate possible effects in great detail, I simply suggest several possibilities for your consideration. First, the tax alone could reduce the developers' demand for vacant land since their demand for land is *derived from* the consumer demand for new housing. If this also results in a lower price for land, the construction companies' marginal cost curve for producing houses will decrease. Counter to this, municipalities participating in the tax program will enter the market for vacant land and thereby increase the demand for land and its price. This will counter the effects of the tax by increasing the marginal cost curve in the housing market. The net effect on the marginal costs of construction companies is not clear, *a priori*. Nor is the effect on real estate agencies clear since the net effect on the price of land may be a decrease, an increase, or no change at all.

Second, the tax on new houses (and/or vacant land) could cause the demand for existing houses to increase. Thus, potential losses by real estate agencies, banks, and insurance companies in the new housing market may be recouped in the market for existing houses.

Third, market demand may actually increase if households believe that the acquisition program will protect the very environmental resources which comprise part of their demand for housing. Again, both construction companies and real estate agencies could actually benefit by the tax program in the long run.

The net effect of all these possible behavioral adjustments in the housing and related markets depends on actual shifts in demand and supply. Furthermore, the net effect on specific construction companies and real estate agencies will depend in part on whether they also own the vacant land.

Finally, consider a demand curve of a different shape—consider the concept, *elasticity*. (This property of demand curves can be difficult to comprehend. Don't fret if you do not grasp it the first or second time.) In Figure 4.2, the effective price increase occurred within the *elastic* re-

gion of the demand curve for houses. By "elastic," we mean that the
percentage change in the quantity of houses demanded,

$$(1,000 - 924)/1,000 = 0.076, \text{ or, } 7.6\%$$

is greater in absolute value than the *percentage* change in total price to
consumers,

$$(\$95,000 - \$96,520)/\$95,000 = -0.016, \text{ or, } -1.6\%$$

Elasticity in this case is

$$e = (7.6\%)/(-1.6\%) = -4.75$$

In other words, a 1% increase in price results in a 4.75% reduction in
quantity demanded in this area of the demand curve.

In general, elasticity is defined as

$$e = \text{(percentage change in quantity demanded)}/\text{(percentage change in price)}.$$

One consequence of a price change within the elastic region of a demand
curve is that total expenditures by consumers decreases (increases) when
price increases (decreases). This is apparent in the example that we just
examined where the price to consumers increased from \$95,000 to \$96,520
but total expenditures (including taxes) decreased from \$95.00 million to
\$89.2 million.

The effect on total consumer expenditures is the obverse, however, if
the price increases within an *inelastic* region of a demand curve. In Figure
4.3, the demand for new houses is drawn much steeper. In this case, the
percentage change in quantity demanded beginning at 1000 houses,

$$(1,000 - 997)/1,000 = 0.003, \text{ or, } 0.3\%$$

is less than the percentage change in price, which is again about 1.6%.
Hence, elasticity is

$$e = (0.3\%)/(-1.6\%) = -0.18$$

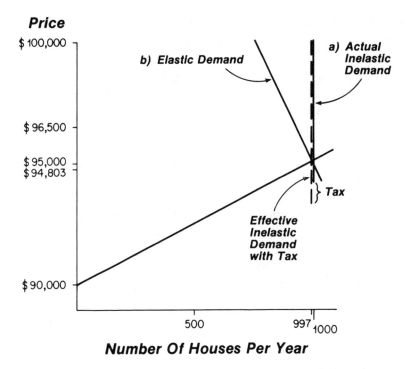

Figure 4.3 Comparison of relatively (a) inelastic and (b) elastic demands.

Although price to the consumers increased as before, total expenditures also increased from $95.0 million to $96.2 million ($96,500 × 997). Thus, within the inelastic region of a demand curve, price and total expenditures vary together.

How does the price increase within the inelastic region of the demand curve in Figure 4.3 affect economic profit? In this example, economic profit is still reduced as before, but by a much smaller amount—a combined loss of $200,000 for *all* construction firms. Also, there is a much smaller deadweight loss. However, consumers lose even more surplus, although most is in the form of the tax transfer. What do you think? Is the demand for property in your coastal zone elastic or inelastic at current prices?

There are other matters to consider as well. In particular, the direct benefits of protecting environmental quality with an acquisition program should be compared to the deadweight losses of the tax. The final result depends on the relative magnitude of potential non-market damages and the effectiveness of acquisition programs.

4. Residents of Petiteport are upset about likely reductions in the quality of municipal services if the town's population increases as projected by a recent growth report. According to a consulting firm that studied the implications of projected growth for the town, it would cost an additional $12 million in property taxes to maintain the same standard of services that current residents are now accustomed to, not to mention the likely additional costs of new services that they do not need now but will need with a significantly larger population. Eight million dollars of these additional costs would be borne by current residents. The firm is somewhat confident that these costs would be distributed as $1.6 million in five years, $2.4 million in 10 years and $4.0 million in 15 years for a total of $8 million. These figures include expected inflation.

The firm made two mistakes. First, the expected costs are not discounted to a present value. Second, the firm used nominal values for costs (including inflation) when it would have been much easier to use real, 1985 dollars. It's okay to use nominal dollars if you also use nominal discount rates, but why bother?

After talking to the consulting firm, we learn that the real, non-inflationary costs are expected to be $0.9 million in 1990, $1.8 million in 1995, and $3.5 million in 2000. Using a 6% discount rate for the town, the present value of these costs is

$$PV = (\$0.9/(1 + 0.06)^5) + (\$1.8/(1 + 0.06)^{10}) + (\$3.5/(1 + 0.06)^{15})$$
$$= \$3.14 \text{ million}$$

Our conversation also revealed that the firm is uncertain of its estimates. In fact, the estimates are "worse case" scenarios. Actually, the firm felt that the probability of these costs coming to fruition is only 30%, although it was very confident of the timing of the costs. More likely (and with a 70% probability), the real costs would be $0.7 million, $1.3 million, and $2.9 million in the respective years.

How do we include this information on probabilities when computing present values? One way is to determine the *expected value* of the costs.

Thus, if the firm thinks that the probability of costs being $0.9 million in five years is 30% (i.e., 0.3) and the probability is 0.7 that the costs will be $0.7 million, the expected value of the costs in five years is

$$(\$0.9 \text{ million} \times 0.3) + (\$0.7 \text{ million} \times 0.7) = \$0.76 \text{ million}$$

Using the same procedure, the expected value of the costs in 10 and 15 years is $1.45 million and $3.08 million, respectively. Therefore, the present value of the *expected* costs is

$$PV = (\$0.76/(1 + 0.06)^5) + (\$1.45/(1 + 0.06)^{10})$$
$$+ (\$3.08/(1 + 0.06)^{15})$$

or, $2.66 million.

This simple analysis reveals how easy it is to misrepresent the expected costs of a project by not discounting future costs to a present value and by mishandling uncertainty about future costs (or benefits).

5. Recreationists in Clamville are concerned that further development around Mercenaria Pond will force the shellfish warden to close the clam beds. They convinced the town to hire a consulting firm to estimate roughly the annual losses in recreational benefits that would result from a closure. The firm was advised by a resource economist at a nearby university to research the contingent valuation literature to get a rough idea of the economic value of recreational shellfishing. After doing so, an employee selected an annual value of $30 per individual. The town's Department of Parks and Recreation told the firm that 1,200 residents went shellfishing last year. Hence, the firm estimated the recreational value to be $36,000 per year.

This is an adequate approach as long as you recognize that you get rough estimates at best. Nevertheless, we still want to clarify a few issues. First, were travel costs deducted from total willingness-to-pay (i.e., does this figure represent consumer surplus or total benefits)? Probably not, but since distances to the pond are likely to be quite short for residents, and the estimate is a ballpark figure anyway, it makes little sense to be picky about this oversight.

There is, however, another potentially important consideration. The employee later ran into the economist at Mercenaria Pond. They were discussing the benefits calculation when the economist asked what study had been used to get the $30 figure. Whoops! There were no close substitutes for recreationists to choose from in the particular study selected by the consulting firm. Yet, residents using Mercenaria Pond have several nearby sites at which to go shellfishing. Hence, the $30 figure could easily be an overestimate of actual benefits. All else being constant, an individual's demand for a shellfishing site is both smaller and more elastic when substitute sites are reasonably available. Consumer surplus is less in this case. The economist also pointed out that the benefit assessment should include residents who are currently non-users but who, nevertheless, have an option value for possible future use. Nor does the figure for use benefits include economic existence values for wildlife and future generations.

    6. It is well known that wetlands, including salt marshes, assimilate nutrients from sewage. In fact, wetlands can provide the equivalent of tertiary treatment. This functional service leads some people to argue that if an acre of saltmarsh can do the same job as a tertiary treatment plant which operates at, say, $25,000 per year, then the acre's economic value in waste assimilation is also $25,000 per year.

This reasoning is fallacious. First, economic value is ultimately bounded by the demand for a commodity, which, in this case, is an environmental service performed by salt marshes. Consequently, if residents do not demand the level of waste treatment that salt marshes can provide, then the claim does not even begin to have a basis in economics (Figure 4.4a).

However, the demand for wetland acreage to perform waste assimilation may actually be growing in many coastal areas which have a groundwater pollution problem or are facing expensive alternatives for sewage disposal in general. Conceivably, the quantity demanded for tertiary treatment may equal or exceed the capacity of an acre of wetlands to assimilate waste (Figure 4.4b). In this case, the operating costs of tertiary treatment (area under the marginal cost curve) would underestimate the total economic value of waste assimilation (area under the demand curve).

    7. The town of Gull Hill learned recently of the possibility of sending municipal trash to a regional disposal facility. Unfortunately, the

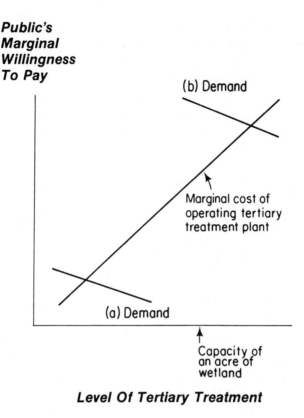

Public's Marginal Willingness To Pay

(b) Demand

↖ Marginal cost of operating tertiary treatment plant

(a) Demand

↑ Capacity of an acre of wetland

Level Of Tertiary Treatment

Figure 4.4 Demand for tertiary treatment of sewage.

town already spent $600,000 on land for a new landfill. As a result, several people are arguing that it is better to continue with the original plans so as not to waste the $600,000.

Although the conclusion that it is "better" (presumably, the supporters of this stance mean "more efficient") to continue with the plans for the landfill may turn out to be correct, the logic is incorrect. The expenditures to date on a landfill site are sunk costs—what's done is done. From an economics standpoint, it would be wiser to assess the cost-effectiveness of the two alternatives starting at the present time: the present value of the future costs of sending trash to the regional disposal facility versus the present value of future costs associated with continuing with the landfill project. Forget about sunk costs.

8. An engineering firm is trying to convince the town council of Aquifer to install a slurry wall up-gradient of a public well. The wall will divert a plume of toxic waste from a nearby naval base that is encroaching upon the well and a nearby pond. The firm argues that not only will the slurry wall protect groundwater quality and the water quality of the nearby pond for recreation, but it will also protect the property values of owners who receive water from the well, thereby adding immensely to total economic benefits.

Although the slurry wall might protect water quality in the well and the pond and protect property values, it would be incorrect to add these benefits together. Adding the independent estimate of avoiding damages to property values to the estimates of the values of protecting potable water and of aquatic recreation would be double counting the benefits of the wall since they measure the same thing.

9. Three people are discussing—indeed, arguing—about the relative costs and benefits of land use strategies which are designed to protect the coastal environment. One person insists that the only thing to do is buy the remaining open space; acquisition is the only surefire way to protect natural resources in a natural state.

    A second person thinks that this idea is quixotic. He argues that towns cannot afford acquisition programs of the size that would be needed. Regardless, property owners never would vote to raise taxes by large amounts. The best alternative, he claims, is to tax all new development at a rate that covers the external costs of new development. That's fair, isn't it? After all, it is new development that will cause the additional problems.

    A third person counters that no elected official would support such a large transfer tax. What is needed, he says, is a combination of a conservation easement program and a transfer-of-development-rights (TDR) program. In this way, landowners and developers could either volunteer for one program or be compensated by the other. That's democratic.

    The others think that he is naive. They claim that an insufficient number of people would participate and important benefits would still be lost. Regardless, the costs of administering a large TDR program would be too high. Furthermore, there aren't enough places in the town to which to detour expected development anyway.

This is certainly a convoluted exchange. The only point that I wish to make is that a careful consideration of the alternatives would frame the discussion. It is important here to keep the with-versus-without distinction in benefit-cost analysis clear. In this case, we have four alternatives: status quo, acquisition, transfer tax, and a combination of conservation easements and TDRs. The benefits and costs of each alternative need to be quantified and expressed in present values. Chapter 10 simulates the type of analysis that is needed.

10. An ardent environmentalist complains that economists have an annoying habit of placing a dollar value on everything. "Economists know the price of everything and the value of nothing," he accuses. We should preserve open space for wildlife. We should bequest a clean environment to future generations. These things should be done regardless of cost.

Well, some economists might monetize "everything," but some economists recognize the possibility of intangibles. And I am not sure whether this person would be willing, or able, to pay all that is necessary to protect the environment for wildlife and for future generations. But he does raise a valid point concerning the purview of economic analysis and the possibility that society—you and I—have objectives other than efficiency in mind when we consider the allocation of coastal resources among mutually exclusive uses.

Hopefully, you have gained an understanding of why economists are able to express certain values in monetary units, even when markets do not exist and when use is not involved. The valuations reflect personal utility and economic profit. Markets and prices are not necessary for well-defined economic values; only a rational interest in one's own well-being and an indifference mapping of preferences such as in Figure 2.12 are needed as a foundation for economic valuations. Economists do not question peoples preferences and valuations; we seek to measure these completely and accurately.

PART TWO: CASE STUDIES

Chapter 5

Demand for Galápagos Tourism: A Case Study

Background

Advertisements in nature magazines and the offerings of travel agencies reveal numerous and varied opportunities for people to vacation in exotic, natural places. Unfamiliar wildlife and panoramic views of landscapes and seascapes lure tourists to African photo-safaris, week-long whale-watching cruises, and even the hostile environs of Antarctica. Whether due to a marked change in preferences or simply to increased opportunities, leisure time, and income, the demand for nature vacations seems to be growing rapidly. This chapter reports on tourist demand for vacations in the Galápagos Islands—one of the most pristine and diverse environments available to tourists.

The Galápagos Islands strattle the equator approximately 600 miles west of Ecuador in the Pacific Ocean. Much of the land area is jagged volcanic rock and is sparsely covered by cacti and other desert flora. In contrast, cool nutrient-rich Pacific waters support a diverse and abundant assemblage of marine life. In addition, the desert climate and productive waters offer tourists highly unusual, if not unique, opportunities to view a large variety of wildlife in one place. Thus, tourists encounter giant tortoises, sea lions, penguins, coral, skip-jacks, and endemic species like the marine iguana (to name a few) as well as a large array of bird species such as the flightless cormorant, waved albatros, and the colorful blue-footed booby.

Ecuador's commitment to the protection of the Galápagos Islands and its biota led to the establishment of the Galápagos National Park in 1959. Tourist activity within the Park, which comprises 90–95% of the archipelago's land area, is strongly regulated by the Park through licensed guides. Damage to the environment is kept to a minimum by restricting visitors to well-marked paths. In addition, because the vast majority of tourists have accommodations on boats, destruction of the environment for land-based services is minimized.

Although the Park is large and tourists are closely regulated, there is growing concern that tourism growth and development is becoming a serious threat to preservation. The number of tourists is now near the official annual limit of 25,000 visitors. There has also been a large increase in the number of vessels offering accommodations and tours, with more entry expected. In addition, a second airport was recently opened to facilitate transfer of tourists to and from the Galápagos Islands, and some entrepreneurs are pushing for permits to construct hotels and other Caribbean-type tourist facilities.

The trend toward increased capital investment and tourism raises questions about conflicts between wildlife preservation and likely "traditional" water activities such as sportfishing, sunbathing, waterskiing, and recreational boating. It also raises doubts about the compatibility of "purist" interests in nature and Caribbean-type vacations which are crowded with people and make heavy use of the environment and its natural resources.

Fortunately, the Galápagos Islands probably exemplifies how tourism and conservation can benefit from each other. First, one may question the wisdom of creating yet another island resort with hotels, restaurants, and water sports in a remote part of the world when there already exists a large number of substitutes which are probably less expensive for most people. Perhaps more important though, is that the nature-loving, Galápagos tourist—the type of tourist with a demonstrated willingness to pay up to several thousand dollars for just a week's stay in the Galápagos—will probably be sensitive to a zoo-type of experience with too many people and buildings. Nevertheless, little is known about the economic demand for the present type of nature-oriented tourism and its potential for attracting badly needed foreign revenue into Ecuador. This chapter presents an analysis of individual demands for vacation days in the Galápagos and aggregates individual demand curves to derive an aggregate tourist demand. See Broadus and Gaines (1987) for background information on Park and marine management in the Galápagos Islands.

The Study

The commodity, a Galápagos vacation, presents several problems for traditional demand analysis. First, the vast majority of tourists visit the

Galápagos only once in their lifetime; hence, it is not possible to estimate an individual demand curve for the number of visits. In addition, there is no such thing as a homogeneous Galápagos vacation. Vacation packages vary considerably in terms of length of stay (2 to 14 days), types of accommodations (a small, 4-to-16 passenger boat with shared toilet and possibly no shower to luxury accommodations with private bath, gourmet meals, and air conditioning on a 90-passenger liner), and price (a range of several thousand dollars depending on duration, accommodations, and nationality of tourist). Consequently, the travel cost technique cannot be used to estimate an aggregate demand equation for visits. Finally, many tourists visit several places in addition to the Galápagos, such as Machu Picchu and the Amazon, during the same vacation. Thus, it is virtually impossible to allocate the cost of the entire vacation package to its various destinations.

Fortunately, the hedonic price technique (see Chapter 3) is suited to the analysis of commodities such as Galápagos vacations which are characterized by several distinct attributes, such as duration, type of accommodations, number of destinations, and so on. The first step is to estimate the hedonic price equation for Galápagos vacations which quantifies the relationship between price and attribute levels. Once estimated, the price of a single attribute that is implied by the equation can be derived. In particular, we will derive the implicit price of a vacation day in the Galápagos. Finally, the implicit price of a vacation day is calculated for each tourist in the sample and combined with data for other relevant variables, such as income, to estimate an individual demand equation for days in the Galápagos (not visits or trips). Using days in the Galápagos rather than number of visits is helpful because the annual visitor quota of about 25,000 is actually based on visitor-days. The belief is that more than 25,000 visitors spending an average of five to six days in the Galápagos would lead to environmental damage.

Two data sets were collected for this analysis. Data describing different Galápagos vacation packages (including possible trips to other destinations in South America) were used to estimate a hedonic price equation and the implicit prices for days in the Galápagos and for days in "other places." This equation and the derivations are merely means to an end— the demand estimation—and, therefore, are not presented here.

The second data set was obtained from questionnaires given to tourists at the end of their vacation. The questionnaire, which was written in four languages (English, Spanish, German, and French), elicited information on the length of the total vacation, number of days spent in the Galápagos, type of accommodations, satisfaction with the wildlife tours, income, and age. These data and the implicit prices for each respondent were used to estimate an individual demand equation for days in the Galápagos.

Results

The following equation is a linear, statistical demand model for Galápagos vacation days, q:

$$q = 3.27 - 0.017 \times P_G + 0.632 \times P_S$$
$$+ 0.0002 \times \text{Income} + 0.044 \times \text{Age}$$

where P_G = implicit price of vacation days in the Galápagos,
          P_S = price of vacation days at other destinations that are part of the vacation package and, therefore, are considered to be substitutes,
    Income = per capita income (in $US), and
      Age = age of the respondent.

Notice that the coefficients on the price and income variables have the expected signs. That is, the choice of number of days spent in the Galápagos decreases as own-price increases, increases as the price of the substitute destinations increases, and increases with income. In addition, the number of days spent in the Galápagos increases with age of the tourist, perhaps reflecting an increase in vacation or leisure time available to older workers and retirees.

The above demand model can be used as a basis for deriving an aggregate, market demand curve for days in the Galápagos. As illustrated for consumers and suppliers of shrimp in Chapter 2, we want to combine the individual demand curves for all tourists. First, we assign values to the explanatory variables other than own-price (i.e., P_G). One reasonable alternative is to use their average values: P_S = $9, Income = $32,000, Age = 50. Substituting these values into the above model yields the individual demand equation:

$$q = 3.27 - 0.017 \times P_G + 0.632 \times 9 + 0.0002 \times 32{,}000 + 0.044 \times 50$$
$$= 11.80 - 0.017 \times P_G$$

Notice that other values for these variables would change the value of the intercept and, therefore, shift the position of the demand curve.

Next, we combine the individual demand curves mathematically using Ecuador's quota of 25,000 annual visitors as the basis for the aggregation. The concatenations illustrated by Figures 2.3 and 2.7 in Chapter 2 can be achieved mathematically by multiplying the above demand equation by 25,000; therefore,

$$Q = 25{,}000 \times q = 25{,}000 \times (11.80 - 0.017 \times P_G)$$
$$= 295{,}000 - 425 \times P_G$$

where Q is aggregate tourist demand. Both the individual and the aggregate demands are illustrated in Figure 5.1.

The aggregate demand curve can be used to examine the potential effects of Ecuador's visitor quota on Galápagos tourism revenue, most of which is "imported" from international tourists. The reader may recall that total revenue increases when price increases within the inelastic region of a demand curve. Thus, one should expect total revenue to be maximized at the point of unitary elasticity. We can calculate this point using a modified version of the elasticity formula,

$$e = -425 \times (P_G/Q)$$

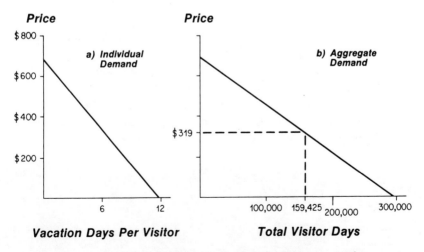

Figure 5.1 Individual and aggregate demands for Galápagos vacations.

where -425 is the coefficient on the own-price variable in the aggregate demand equation. Conceptually, the own-price coefficient in a linear model is the change in quantity demanded divided by the change in price. The price corresponding to unitary elasticity is solved by substituting the equation for Q into the elasticity formula and setting the result equal to -1.0. In this case, $P_G = \$347$ and $Q = 147,500$ total visitor days. In contrast, the average implicit price of vacation days calculated for respondents to the survey is \$319, and the number of total visitor days corresponding to this price is 159,425. This quantity is probably not significantly different from 147,500 given the assumptions built into the calculation. Thus, it appears that, by minimizing damage to the natural environment with a quota, the government is maximizing the revenue from Galápagos vacations.

Concluding Remarks

This chapter demonstrates how market data for vacations can be used to estimate tourism demand curves for natural environments such as the Galápagos Islands. The interaction between conservation and tourism demand was examined for the single issue of revenue maximization. Of course, the Ecuadorian government could capitalize on the dependency of tourism on conservation in other ways, although each option will have distributive implications. For example, the government could maximize tax revenue from entrance fees charged to Park visitors. Using the above aggregate demand model, the optimal tax is about \$173.50 *per visitor-day*. (Tourists from Ecuador are now charged \$6 for the *entire visit,* while foreigners are charged \$40.) This would generate approximately \$12 million US for Ecuador's treasury that could be spent on national programs including preservation of the Park. However, revenues to the tourism industry would decline substantially by about \$25 million US. In addition, many fewer people would visit the Park due to the higher total price.

Chapter 6

Potential Economic Effects of Relative Sea Level Rise on Bangladesh's Economy: A Case Study

Background

Many scientists agree that sea level is once again rising and perhaps at an increasing rate. For example, Hoffman (1984) projects that the "greenhouse effect" will result in a 1.8 to 11.3 foot increase in global sea level by the year 2100. His projections are based on expected increases in atmospheric temperatures that, in turn, would increase the volume of the oceans as ice masses melt and through thermal expansion of sea water. In order to put these projections into perspective, consider the possibility that Florida and Louisiana could lose about 20% of their land area to sea level rise. (For starters, see Barth and Titus's [1984] book on sea level rise.)

In addition to global sea level rise, several other processes and activities threaten to inundate our coasts. In particular, coastal land is sinking relative to sea level in many coastal areas. Often the sinking is due to tectonic movements of the earth's crust and to subsidence, or the protracted rebound of the earth's crust following the most recent ice age. However, sinking can also be induced by the withdrawal of groundwater or petroleum and gas resources. In addition to these factors, sea level will rise relative to land level near the mouths of dammed rivers. In this case, the supply of sediment to the coast is interrupted, and losses due to erosion are uncompensated. Together or individually, these factors exacerbate expected increases in global sea level.

The potential physical effects of increases in relative sea level need to be evaluated in terms of their potential economic implications. Certainly, concern is heightened by the fact that much of a coastal nation's economic base and population occurs near the coast. In a qualitative sense, the expected increases will inundate massive areas of residential, commercial, and industrial land and immobile capital (e.g., buildings) and displace millions of people worldwide. In addition, beach recreation and tourism are threatened by accelerated coastal erosion. Coastal flooding and storm damage may be exacerbated, particularly where storms intensify in strength and frequency. Finally, the salinity of groundwater and estuaries could significantly reduce the size of potable aquifers and valued fish and shellfish populations. Oceanic islands and countries with large river deltas, such as the Ganges–Brahmaputra–Meghna Rivers Delta in Bangladesh, are especially threatened. This chapter reports on a study of the economic importance of land potentially exposed to sea level rise in Bangladesh.

The Study

Bangladesh is a densely populated, agrarian nation in southeast Asia (Figure 6.1). In 1985, its total Gross Domestic Product (GDP) was less than $16 billion US (i.e., 406 billion Takas), about 50% of which was directly attributable to agriculture. Not surprisingly, industrial and commercial activity tends to occur near and within large cities along the major rivers. Severe storms which emanate from the Bay of Bengal flood low-lying areas annually, revealing the substantial area of land in Bangladesh that is susceptible to sea level rise. Figure 6.1 illustrates two possible scenarios for relative sea level rise. The 0.8 meter line (i.e., 2.7 feet) corresponds to natural subsidence and a conservative increase in global sea level. The more extreme 3.4 meter line (i.e., 11.1 foot) incorporates subsidence, global sea level rise, and the possible effects of damming rivers.

Assessing the economic impacts of future sea level rise is considerably more difficult in practice than in theory. Part of the problem is attributable to the enormous scientific uncertainty surrounding projections of future global climate and the resultant effects on sea level. This problem is compounded in countries like Bangladesh where detailed information on economic activities, assets, and resources are not readily available. Researchers in the United States favor the use of market values of land and immobile capital assets such as buildings since it can be argued that these data approximate the present value of a stream of expected annual profits. However, even these studies overlook both losses in consumer surplus

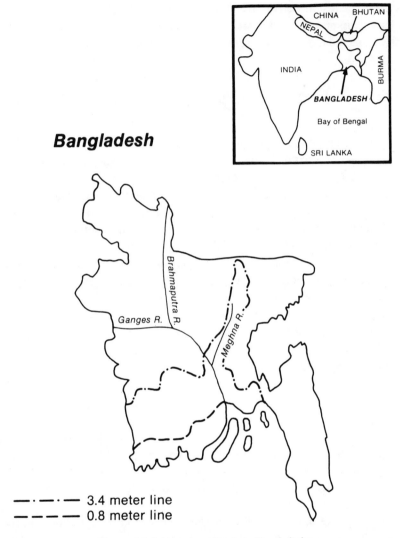

Bangladesh

—·—·— 3.4 meter line
— — — — 0.8 meter line

Figure 6.1 Projected sea levels in Bangladesh.

due to reduced production and non-market damages attributable to salt-water intrusion. Because of these difficulties, national data for GDP in Bangladesh were used as a basis for a first order approximation of the level of economic activity occuring within the 0.8 meter and 3.4 meter zones. The results derived from the GDP data serve as an adequate gauge of the market activity potentially affected by relative sea level rise but, as will be seen, do not themselves constitute estimates of potential losses.

Results

The economic activities originating within the 0.8 meter and 3.4 meter zones were approximated on the basis of GDP statistics reported by the Economic Intelligence Unit (*Quarterly Economic Review of Bangladesh*) and indexed to percentages for land area, population, and industrial activity potentially exposed to relative sea level rise. For example, consider the calculations for GDP activities indexed to the distribution of the population (Table 6.1). Total GDP for these activities is about $2.92 billion US (or about 74 billion Takas). Since about 5% of the population lives within the 0.8 meter zone, it was assumed that approximately 5% of population-based production and services, or $0.15 billion US, would be affected by a 0.8 meter rise in relative sea level. Table 6.1 summarizes the results of economic activity originating within the sea level rise zones.

The next step is to convert the approximations for total affected economic activity into present values. First, recognize that relative sea level rise is a gradual, cumulative process. Consequently, only a small percentage of the economic activity will be affected during the first year, and this effect will be augmented incrementally each year. For example, geologists contributing to this study suggest that the 0.8 meter rise could occur by the year 2050. Using 1987 as the base year, one-sixty-third (1/63) of the potentially affected economic activity—$11.9 million US—would take place during the first year, approximately two-sixty-thirds (2/63) during 1988, and so on.

Actually, we must allow for economic growth into the future but also discount the future values to present values. To do so, it was assumed that the GDP will expand at the same rates that the World Bank projects for population growth. In turn, these projections of future economic activity affected by relative sea level rise were discounted to present values. Combining these effects, the real value of economic activity within the 0.8 meter zone that is affected t years in the future is:

$$\text{Real value} = (t/63) \times \$750 \text{ millionUS} \times (1 + g)^t$$

Table 6.1

Gross Domestic Product (GDP) Potentially Affected by Relative Sea Level
Rise in Bangladesh ($ billion US)[a]

| | | Scenario for Relative Sea Level Rise | | | |
| | | 0.8 meter rise by 2050 | | 3.4 meter rise by 2100 | |
GDP Account	GDP in 1987/88[b]	GDP	Percent of Total	GDP	Percent of Total
Area based					
agriculture	7.74	0.54	7	2.01	26
Population based					
construction; power, water and sanitation; housing; public adminsitration and defense	2.92	0.15	5	0.57	27
Industry based					
industry; transportation, storage, and communication; trade services; banking and insurance; professional and miscellaneous services	5.82	0.06	1	0.99	17
Overall	16.48	0.75	4.6	3.57	21.7

[a] National GDP information and accounts are reported in the *Quarterly Review of Bangladesh* (1985), The Economic Publishers Ltd.
[b] GDP data for 1984/85 reported by the World Bank were increased to 1987/88 using World Bank projections for population growth rates reported in *World Population Projections 1985*, John Hopkins University Press.

where g is the growth rate of the economy, whereas the corresponding present value is:

$$\text{Present value} = \text{Real value} \times (1 + r)^{-t}$$

where r is the social discount rate. A similar procedure was followed for the time interval 2050 to 2100, during which time relative sea level rise is projected to increase more rapidly. The present values for each future year were then added to approximate the total present value of GDP activity potentially affected by a rise in relative sea level.

Figure 6.2 illustrates the effect of the social discount rate on the total present value estimates. The present values of potentially affected GDP decrease from $81.1 billion US at a 1% discount rate to about $890 million US at a 15% discount rate. Although it's not possible to know the correct social discount rate, one of the higher ones might be most representative if the low standard of living in Bangladesh results in strong preferences for present consumption.

These crude approximations suggest that relative sea level rise could affect a sizeable portion of the Bangladesh economy in the future. However, we should try to interpret the welfare implications of these GDP-based approximations because decisions to avert future sea level rise problems tend to require very expensive capital investments. Any such

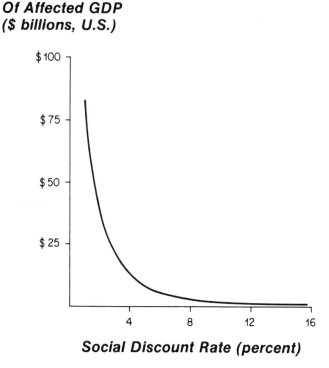

Figure 6.2 Present value of gross Domestic Product potentially affected by relative sea level rise.

investments need to be based in part on the net benefits to Bangladesh. Indeed it would be unwise to invest publicly in sea level control projects if the monies would yield greater benefits in the area of food production and health projects.

Two issues may come to mind. First, GPD statistics are, in effect, derived from total revenue data. Therefore, estimates derived from them exclude consumer surplus and overestimate economic profit. Second, we would like to approximate the welfare effects of relative sea level rise following the "without-project" and "with-project" mode of thought as discussed in Chapters 3 and 10. In this case, the "with-project" scenario corresponds to no increase in relative sea level, while the "without-project" case is represented by the economy affected by relative sea level rise. The question becomes, what is the relationship between GDP originating within the 3.4 meter zone and changes in economic surplus (consumer surplus and profit) due to relative sea level rise?

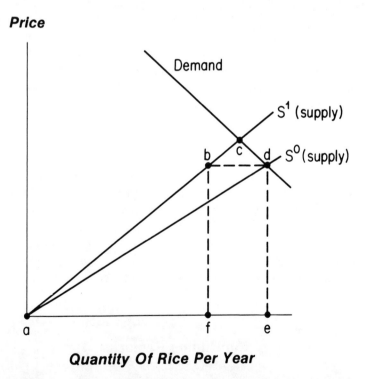

Figure 6.3 Stylized changes in economic surplus due to a rise in relative sea level.

Figure 6.3 helps us to respond to this question, even if only in a very general way. Pretend that we are examining the rice industry, although we could possibly examine more aggregated commodities such as all crops or perhaps domestic output. S^0 is the aggregate supply curve corresponding to the without-sea level rise case. It is the sum of marginal cost curves for all farms. In contrast, S^1 is the supply curve corresponding to sea level rise. S^1 reflects a 26% reduction in the number of farms in the rice industry due to the projected 3.4 meter rise in relative sea level and subsequent loss of agricultural land.

Area bdef represents the part of the GDP account for rice that originates within the 3.4 meter zone. It corresponds to a 26% reduction in farmland used for rice production. However, we are specifically interested in the unknown loss in surplus due to the 3.4 meter rise. By the year 2100, the assumed loss in economic benefits corresponds to area acd. (Notice that the higher price for rice corresponding to point C would prevent a reduction in rice production to point f.) This area is composed of a loss in consumer surplus and a loss in *economic rent*. The latter differs from economic profit because it includes fixed costs. That is, farmers put out of business by sea level rise would still incur fixed costs, and this should be included as part of the potential damages.

The economic damage depicted in Figure 6.3 is about 63% of the GDP estimate. Of course, this comparison is arbitrary since the demand and supply curves are hypothetical, being used only for illustrative purposes. The damage estimate could vary considerably depending on the elasticity of demand and on the position of the supply curves. However, Figure 6.3 does suggest an alternative to using property value data to estimate potential damages attributable to sea level rise once it's determined that more thorough research is warranted. Specifically, researchers could estimate possible reductions in consumer surplus from demand curves and add estimates of fixed costs for displaced firms. One drawback of this approach is that economic profit is assumed to be zero (i.e., the economy is assumed to be in competitive equilibrium). Each approach has difficulties.

Concluding Remarks

The GDP approach uses highly aggregated information to characterize the current scale of economic activity originating within areas expected to be transgressed by relative sea level rise. Although other approaches which measure damages directly from property values are conceptually superior, the GDP approach provides useful first-order information as long as the results are obtained in a logical and sound fashion and are interpreted correctly.

Each approach—property value, GDP, consumer surplus/fixed cost—is associated with measurement errors and errors due to omission. In addition to those already mentioned, each approach omits potential economic damages due to saltwater intrusion, storm damage, and subsistence farming and fishing. A more comprehensive approach to the problem should compare the results of the three approaches, include the additional environmental damages, and further compare the calculations of potential damages to the costs of strategies designed to avert or mitigate losses due to relative sea level rise.

Chapter 7

Household Demand for Local Public Beaches: A Case Study

Background

Mankind is intensely interested in the recreational opportunities afforded by ocean beaches. Our collective interests seem to be dominated by swimming and sunbathing but also include fishing, surfing, SCUBA diving, windboarding, and certainly strolling. Usage ranges from daily trips to a local town beach to week long vacations to distant island resorts. Indeed, coastal tourism and real estate markets are derived to a large extent by demands for beach recreation.

Several factors have in the past and continue to fuel aggregate demands for beach recreation. (See Ducsik [1974] for a historical perspective on beach demand and for a broad discussion of beach management issues.) First, coastal economies attract and employ a sizeable fraction of a coastal nation's work force. For example, the U.S. Bureau of the Census projects that 75% of the U.S. population will live within 50 miles of the coastline by the year 1990. The population alone swells aggregate demands for beaches. Second, expanding transportation systems reduce travel time to beaches—a savings to recreationists. Finally, increases in disposable income and leisure time stimulate demand for beaches. In effect, more income and time increase our ability to recreate. All things considered, there has been a surge in the demand for local beaches.

The diverse and growing demands for beaches confront several difficulties. Several of these difficulties fall within the rubric "multiple-use conflicts." For example, surfing and swimming are incompatible uses of contiguous waters, as are sunbathing and driving four-wheel drive vehicles along a beach. Another form of use conflict pits private and public ownership against each other. There is a tendency for real estate markets to sequester beaches either directly in jurisdictions where private ownership extends to the low tide line or more subtly by not providing essential, complementary services such as parking and access. In this case, public good characteristics of beaches make it difficult to organize the demands of recreationists in order to compete with private demands.

Fundamental to the above problems is the limited supply of public beaches which are suitable for recreation. Nationwide, only a small fraction of the shoreline—certainly less than 5%—is available and suitable for public recreation. Furthermore, even this small fraction is threatened by erosion and sea level rise. In addition to reducing the size of public beaches, erosion and sea level rise create crowding externalities and thereby reduce the value of beaches to recreationists.

In principal, the above problems can be redressed by beach acquisition and erosion control management, although these actions can cost well over a million dollars for even small, local beaches. Not surprisingly, the more effective, long-term options such as fee-simple purchase of land or development rights and beach nourishment are the most expensive. What seems to plague resource managers, then, is whether these costs can be justified in terms of recreational benefits.

This chapter is intended to shed some light on the possible magnitude of beach recreation benefits. The study is an application of hedonic demand analysis as described previously in Chapters 3 and 5. In this case, property values in southern Rhode Island were used to infer the value that households place on local public beaches.

The Study

Southern Rhode Island is endowed with long stretches of sandy, ocean beaches. These beaches are used extensively by year-round residents, by households with summer homes in coastal towns, and by thousands of tourists and "day-trippers" from Massachusetts, Connecticut, and elsewhere in Rhode Island. Not surprisingly, proximity to a public beach is an important determinant of property value. Everything else held constant, property values (and rental values) increase markedly as distance to a public beach decreases.

Looked at another way, however, households can reduce or save on property costs by increasing their distance to the nearest public beach. Nevertheless, savings in the housing market is associated with an important opportunity cost that also involves distance. Specifically, increasing distance also increases travel costs to a beach and makes visiting the beach less convenient. These factors will reduce the number of household visits and, therefore, the benefits of beach recreation. Figure 7.1 illustrates the case for an increase in distance to the beach. Travel cost, or the "price' of a visit increases commensurately from TC^0 to TC^1, causing seasonal visits to decrease from V^0 to V^1. Total benefits are reduced by area ABV^0V^1. More importantly, the increase in travel cost and inconvenience results in a loss in consumer surplus equivalent to area TC^1ABTC^0.

The tradeoff between reduced property costs and losses in consumer surplus presents an opportunity for estimating demands for local public beaches from property values. Conceptually, a rational household maximizes benefits from beach recreation when savings in the real estate market due to a small or marginal increase in distance from the beach is just equal to the concomitant marginal loss in consumer surplus. This occurs at point A in Figure 7.2a. At shorter distances, savings are greater than losses; at farther distances, losses are greater than savings.

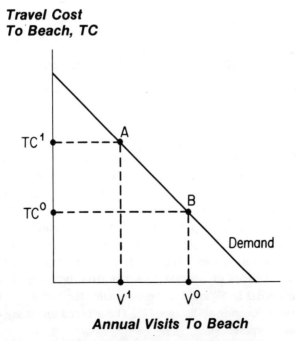

Figure 7.1 Increased travel costs reduce surplus from beach use.

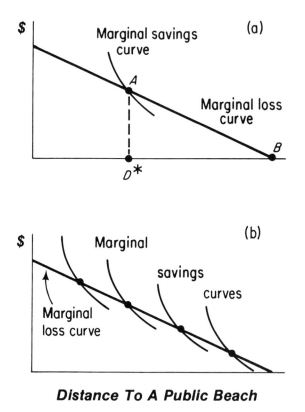

Figure 7.2 Property markets reveal benefits of nearby public beaches: (a) household equilibrium and (b) identifying a marginal loss curve.

Actually, property value analysis does not allow estimation of visit demand curves per se. However, it does allow us to estimate consumer surplus indirectly. Specifically, shifts in the curve for marginal savings which are apparent from an analysis of property values will trace or identify the loss curve as in Figure 7.2b. Although the mathematics is beyond the scope of this introductory book, it can be shown that the area below a household's marginal loss curve and to the right of its distance to a beach is an estimate of consumer surplus from beach recreation. Intuitively, area ABD in Figure 7.2a approximates the losses that the household *avoids* by locating at distance D*. The avoided losses are equivalent to consumer surplus.

The following estimates of loss curves and consumer surplus are based on the analysis of 738 properties sold in South Kingstown, Rhode Island, between 1979 and 1981. A linear version of the hedonic price model for coastal property from which marginal losses were estimated is reported by Anderson and Edwards (1986). See their paper for more complete details.

Results

The estimated marginal loss equation is:

Marginal loss = 2,837 − 225 × Distance + 0.031 × Income

where Distance is distance to the nearest public beach and Income is annual household income. Figure 7.3 illustrates the equation for a household with a $25,000 income (1980 dollars). Marginal loss decreases from about $3,556 at one-quarter mile from the beach to about $237 at 15 miles from the beach.

Let's suppose that the household's property is four miles from the beach. Consumer surplus is approximated by area ABD, or $16,340. This figure should be interpreted carefully, however. First, this is not an approximation of annual consumer surplus. Rather, it approximates the present value of a time series of expected annual consumer surpluses for the household. Using 30 years as the terminal time period and a 5% rate of discount, the annualized value, or annuity, is about $1,063. In general, annuities (A) can be estimated from present values (PV) by the formula:

$$A = PV/([1 - 1/(1 + r)^T]/r)$$

where r is the discount rate (e.g., 0.05) and T is the terminal time period (e.g., 30 years). In effect, this formula converts a present value into an annualized stream of equal-sized benefits.

Second, the approximation is on a household basis. Assuming that there is an average of four recreationists in the household, annual consumer surplus per person is approximated to be $266. The reader can use the loss equation and the conversion formula to approximate consumer surplus for households with other incomes and locations. Maintaining the

**Marginal Loss
In Consumer
Surplus**

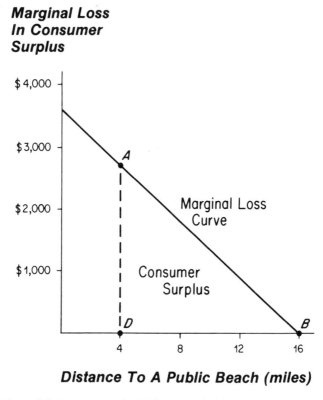

Distance To A Public Beach (miles)

Figure 7.3 Consumer surplus behind a marginal loss curve for distance.

above assumptions, Figure 7.4 illustrates approximations of annual consumer surplus per capita for several other cases.

Before ending this section, it is instructive to compare the possible magnitude of recreational benefits with the cost of beach management. As a realistic, albeit hypothetical, case, consider a small coastal community of 12,000 households and with two miles of public beaches. Years of erosion reduced the size of the beaches considerably. Let a generous estimate of the present value of costs for a 50-year beach nourishment and maintenance project be $1.5 million per mile or $3 million total. What are the recreational benefits to these households of erosion control? Assuming that consumer surplus for the average household is similar to that depicted in Figure 7.3, a first order approximation of the present value of benefits of this project is 12,000 households times $16,340 = $196 million.

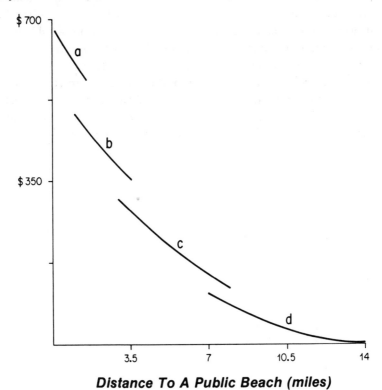

Figure 7.4 Effect of distance and income on surplus from beach use. Incomes are (a) $60,000, (b) $35,000, (c) $25,000, and (d) $15,000.

The benefit approximation dwarfs the cost figure. Even if we allow for substitution with visits to beaches in other towns—a possibility that is not certain given the indiscriminant impacts of erosion and relative sea level rise—erosion control for our hypothetical community is probably an efficient public investment.

Concluding Remarks

This study only begins to explore the economics of beach management. There are a few additional studies in the economics literature, but much more work needs to be done before we can evaluate the efficiency of public investment into beach acquisition and erosion control. On the benefits side of the equation, we need to learn much more about local user demands as well as demands by tourists and other households who do not reside in beach communities but who live close enough to make daily trips. With this information at hand, we can determine the efficiency of beach acquisition and erosion control projects. Furthermore, we can assess economic tradeoffs between private and public ownership of beaches and determine whether it is possible to compensate land owners and still achieve net economic gains—the Pareto criterion at work.

Chapter 8

Protecting Water Quality in Coastal Lagoons for Recreation: A Case Study

Background

Coastal states along the Atlantic Ocean and Gulf of Mexico in the United States are fringed with coastal lagoons. Many people are moving to these areas in part because the lagoons and nearby beaches offer excellent recreational opportunities including swimming, fishing, shellfishing, boating, and bird watching. Ironically, though, the collective use of the coastal environment is ruining the very environmental resources that people come to enjoy. An important set of problems centers on sewage leachate that is transported via groundwater flow into coastal lagoons, thereby reducing water quality and threatening recreational benefits.

Rhode Island shares the pollution problems now faced by many coastal states. Between 1940 and 1980, the number of houses in southern Rhode Island increased elevenfold. Also during this time, many summer cottages were converted to year-round residences. This burgeoning residential growth, almost exclusively in unsewered areas, has contaminated parts of several lagoons with sewage pathogens. Fecal coliform standards for safe shellfishing and, in some cases, for swimming are now exceeded during the summer in parts of several lagoons. More widespread contamination and eutrophication are likely in the near future because there is still room for a tripling of the number of houses near the lagoons under current zoning regulations.

The Study

This case study assessed certain economic impacts associated with the water pollution problem in southern Rhode Island. The study centered on the recommendation promulgated by the State's Coastal Resources Management Council that vacant land within the coastal lagoons region should be downzoned from two-acre zoning to five-acre zoning per buildable lot. The town of South Kingstown was selected for a case study since it was ready to consider the downzoning recommendation. Approximately, 2,890 acres of land enveloping 118 privately owned parcels would be affected. The maximum number of buildable lots would decrease from 1,319 to 386. Only basic results of the analysis are presented here; see Anderson and Edwards (1986) for more details.

The economic impact analysis considered effects on only two groups in the town: owners of vacant land and current and potential recreationists in the town. If a landowner has a 10-acre undeveloped parcel, the market value of the parcel would be higher under two-acre zoning than under five-acre zoning; "extra" land is not valued as highly as land needed for the house's foundation and the driveway and to satisfy setback requirements. Also, it might be more difficult to sell the larger, more expensive five-acre lots. Thus, in present value terms, the land's market value would be expected to fall under downzoning.

However, as water quality declines, recreationists would either continue to use the lagoons, deriving less satisfaction as the quality of the site decreases, or switch to a less preferred substitute site. In either case, users' recreational benefits would decrease to a greater extent under current zoning. The downzoning proposal is expected to result in less deterioration in water quality than if development occurred under current zoning.

Results

Losses to Owners of Undeveloped Land

The hedonic price technique was used to predict the expected impacts on property owners of increases in the minimum size of a buildable lot. Water frontage and a view of a coastal lagoon or the ocean and distance to these water bodies were significant determinants of property values (Table 8.1). For example, a typical three-bedroom house on two acres of land situated on a coastal lagoon and one mile from an ocean beach had an estimated

Table 8.1

Hedonic Price Equation for Residential Property in Southern Rhode Island[a]

Price of property = 22,156 + 0.30 × square footage of lot
+ 7.7 × square footage of house
+ 9,755 × number of bathrooms
− 127 × age of house in years
+ 7.5 × square footage of porch
+ 14.9 × square footage of garage
+ 9.7 × square footage of finished basement
+ 5,621 × number of fireplaces
− 186 × miles to the university
− 630 × miles to the nearest public beach
− 741 × miles to the nearest coastal lagoon
+ 166 × feet of frontage on salt water body
+ 3,101 × year of sale[b]
+ 11,156 × view of salt water body[c]
+ 1,012 × location in wooded area[c]

[a] See Edwards (1984) for additional models including the optimal functional form.
[b] 1979 = 1; 1980 = 2; 1981 = 3
[c] These variables are equal to 1 if the condition is true and 0 if false.

value of $168,400 (1980 dollars). For the sake of comparison, an identical house five miles from the nearest ocean beach and four miles from the nearest coastal lagoon would be valued at $76,305 in the same market.

We assumed that 10 of the 118 parcels would be sold each year after the property had been divided into five-acre lots. This rate of development is somewhat higher than the average rate over the previous six years but is representative of "boom" development observed in 1980. The total discounted values of these vacant parcels were predicted using the estimated hedonic price model for vacant land under two-acre and five-acre zoning. The net present value of losses to owners of downzoned parcels was estimated to be about $10 million using a 6% social discounting rate (Table 8.2). However, this figure probably overestimates potential losses since (1) the reduction in the availability of building lots would actually increase price over time, and (2) there was no account of the potential protection of property values due to the protection of water quality. An alternative development scenario, "slower development," attempting to take these factors into consideration, resulted in a lower estimated loss of $6.4 million.

Table 8.2
Net Present Value (NPV) of Selected Economic Impacts of a Proposed
Downzoning Program for South Kingstown, RI.

Economic Impact	Discount Rate (dollar amounts in millions)			
	0.00	0.02	0.04	0.06
Present value for each impact:				
1) Loss to landowners				
a) accelerated development	−$32.39	−$20.28	−$13.84	−$10.09
b) slower development	−20.54	−12.86	−8.77	−6.40
2) Gains to recreationists				
a) shellfishing & swimming	19.26	12.08	8.22	6.00
b) shellfishing, swimming,				
fishing, etc.	29.08	18.24	12.42	9.07

Net present value of the four possible combinations of the above impacts.
3) accelerated development scenario with protection
of shellfishing and swimming

	−13.13	−8.20	−5.67	−4.09

4) accelerated development scenario with protection
of shellfishing, swimming, fishing, etc.

	−3.31	−2.04	−1.42	−0.66

5) slow development scenario with protection
of shellfishing and swimming

	−1.32	−0.78	−0.55	−0.46

6) slow development scenario with protection
of shellfishing, swimming, fishing, etc.

	9.74	5.38	3.65	3.03

Recreational Benefits

The contingent valuation technique was used to survey residents' willingness-to-pay (1) to protect shellfishing, (2) to protect swimming and activities other than shellfishing that nevertheless require physical contact with water, and (3) to protect activities in category (2) plus fishing, bird watching, and those other activities threatened by the demise of the lagoons. On average, the respondents to the survey bid $77/year to protect shellfishing, $87/year to protect swimming and other water-contact activities, and $169/year to protect all types of recreation. Frequency of use, income, property ownership in the coastal lagoons region, and uncertainty about future demand were statistically significant determinants of willingnes-to-pay in the logarithmic models (Table 8.3).

These results (adjusted for several sources of biases) were used to estimate the total benefits of protecting water quality for residents and non-resident property owners. Since bids are on an annual basis, bids for future years were discounted to present values. Estimates of present values range from about $6 million to $29 million, depending on the change in water quality being valued (categories 1 plus 2 above and categories 1 plus 3 above), and the discount rate (Table 8.2).

Net Present Value of Selected Impacts

The net present values for only the two impacts are also listed in Table 8.2. The estimated losses to landowners under the accelerated development scenario exceed estimates of gains to recreationists for each of the

Table 8.3
Household Willingness-to-pay Models for Protecting Water Quality in Coastal Lagoons. Each variable is in natural log form.[a]

1) Model for protecting opportunities for shellfishing and swimming

Willingness-to-pay = $-4.80 + 0.200 \times$ number of annual visits
$+ 0.618 \times$ feet of frontage on coastal lagoon
$+ 0.756 \times$ annual household income
$+ 0.362 \times$ ownership of other property near coastal lagoons[b]
$- 0.362 \times$ age 70 or older[b]
$- 0.441 \times$ uncertain about future demand[b]

2) Model for protecting all recreational opportunities, including shellfishing, swimming and fishing

Willingness-to-pay = $-5.64 + 0.139 \times$ number of annual visits
$+ 0.525 \times$ feet of frontage on coastal lagoon
$+ 0.903 \times$ annual household income
$+ 0.335 \times$ ownership of other property near coastal lagoons[b]
$+ 0.362 \times$ view of coastal lagoon from property[b]
$- 0.590 \times$ age 70 or older[b]
$- 0.397 \times$ uncertain about future demand[b]

[a] See Anderson and Edwards (1986) for more details.
[b] These variables are equal to 1 if the condition is true and 0 if false.

changes in water quality. However, we get the opposite result for the scenario where future development proceeds at a slower rate and the entire health of the lagoons are threatened.

A definitive statement cannot be made about the efficiency of the downzoning program. Hydrogeological and oceanographic information on the exact relationship between housing development and water quality in the lagoons is not known. Furthermore, other direct and indirect effects of downzoning were not included (e.g., *net* multiplier effects in the construction-related industry, recreational benefits for non-residents, benefits of groundwater protection, damages due to increased traffic, and increased costs of town services). However, the study revealed a comparability between the non-market benefits of protecting water quality and the expected losses to owners of coastal real estate.

Postmortem

In the spring of 1984, a modified five-acre zoning ordinance was adopted by South Kingstown. Because of concerns about the legality of downzoning and of windfall losses for landowners, five-acre zoning was implemented only where development would adversely affect the town's aquifers and not throughout the coastal region. Consequently, only approximately 900 of the 2,890 acres of undeveloped land in the region were finally downzoned. As a result, water quality in the lagoons is expected to continue to deteriorate as has been observed recently in Cards Pond.

In retrospect, the economic analysis provided valuable input to South Kingstown's land use planning process, although non-economic considerations of the distributive effects on landowners prevailed over efficiency goals. Nevertheless, open space preservation continues to be a critical concern in South Kingstown, Rhode Island.

Acquisition programs have been recommended as the most effective and politically feasible alternatives for protecting environmental resources that yield public benefits. Although programs like downzoning might yield a *potential* Pareto improvement in the allocation of resources, the concentration of costs on relatively few individuals with vested interests will tend to undermine such programs. Outright acquisition, on the other hand, results in an *actual* Pareto improvement when residents are willing and able to pay the necessary costs to acquire large amounts of open space. In this case, landowners would receive payment for their land, and towns as a whole would protect the lagoons for recreation.

Chapter 9

Willingness-to-Pay for Potable Groundwater: A Case Study

Background

One can hardly overstate the importance of clean water for sustaining life and economic growth. Of particular interest to this chapter is mankind's increasing dependence on groundwater. (See the U.S. Geological Survey's [1985] discussion of groundwater use and nitrate contamination for additional background information.) For example, groundwater is used by roughly half of the U.S. population for drinking and cooking. In fact, groundwater is the source of drinking water for 97% of rural domestic households and is the sole source of available drinking water for some coastal areas like Cape Cod, Massachusetts. In addition, groundwater is heavily used by industry and the agricultural sector, providing 40% of irrigation use. Furthermore, groundwater helps sustain many aquatic and wetland ecosystems which are valued for fish production, wildlife habitat, and recreation. Finally, an adequate groundwater supply helps prevent saltwater intrusion into potable coastal aquifers.

For the most part, groundwater quality in coastal aquifers is quite good. However, there are numerous isolated instances of contamination with toxic substances such as gasoline. Perhaps more widespread, though, is non-point source contamination of groundwater with nitrate and sodium. Nitrate is derived from the sewage of a rapidly growing coastal population and is probably the most common groundwater contaminant. In addition, saltwater intrusion can contaminate coastal aquifers with sodium.

On-site disposal of household sewage through septic systems and cesspools is a primary source of nitrate contamination in the coastal zone. For example, on Long Island, New York, nitrate concentrations in the upper aquifer averaged about six parts per million (ppm) but exceeded 22 ppm in several test wells. These levels are contrasted to the U.S.

Environmental Protection Agency's standard of 10 ppm for drinking water. Furthermore, nitrate levels were found to be increasing through time. Similar trends are apparent in certain coastal aquifers in Boston, Los Angeles, Delaware, Connecticut, and Dade County, Florida.

Nitrate is believed to be a health hazard if consumed at concentrations above the U.S. Environmental Protection Agency's safety standard. The most clear health effect is methemoglobinemia, or "blue baby disease." This disease is characterized by a reduction in the supply of oxygen to blood cells, thus posing the danger of suffocation. Although less clear, other suspected health risks associated with nitrate include stomach cancer, birth defects, and impairment of the nervous system.

Resource managers and groundwater specialists tend to recommend aquifer protection as opposed to cleanup. On purely cost grounds, it is generally felt that remedial action is more costly than steps necessary to avert contamination in the first place. In addition, cleanup can take many years and may not completely solve the problem. However, future nitrate contamination of otherwise clean aquifers is inherently uncertain. Actually, it is not clear *a priori* that the economic benefits of groundwater protection are greater than the costs of averting uncertain contamination.

Uncertainty about future nitrate contamination of a coastal aquifer has at least two important implications for economic analysis of aquifer protection. First, the present value of the cost of mitigating uncertain future contamination is less than the cost of mitigating certain future contamination. This was illustrated by argument 4 in Chapter 4 where the notion of expected value was introduced. In addition, a person's maximum willingness-to-pay to protect groundwater quality will likely be affected by the probability of future supply. That is, a person is likely to be willing to pay less for groundwater protection when it is not clear whether the aquifer will be contaminated with nitrate in the future. Conversely, willingness-to-pay should increase as the likelihood of future contamination increases.

This chapter reports some of the results of a study of the economic benefits of protecting Cape Cod's "sole source aquifer" from nitrate contamination. It illustrates the use of the contingent valuation method to elicit values of non-marketed resources from households. Although municipalities sell water to residents, water prices probably do not reflect value related to health, property value, the convenience of using tap water, and interests in protecting groundwater for use by future generations. The contingent valuation method facilitates analysis of resource values and of the effects of supply uncertainty on resource valuations.

The Study

Cape Cod's sole source aquifer is virtually the only direct source of drinking water for its inhabitants. Although nitrate contamination of the public water supplies throughout Cape Cod is not imminent, there is a clear trend of increasing nitrate concentrations in many of the public wells. This trend appears to be caused by on-site sewage disposal by the rapidly growing population of year-round and seasonal residents. More than 90% of the Cape's households use septic tanks or cesspools to dispose of household wastes.

During the summer of 1986, a sample of 1,000 households in the town of Falmouth received a mail questionnaire concerning the valuation of their potable supply of groundwater. One section of the questionnaire described a hypothetical scenario for future nitrate contamination and a referendum for averting nitrate contamination. Respondents were instructed to vote either "yes" or "no" as to their willingness-to-pay the amount entered in their questionnaire. The amounts ranged from $10 to $2,000 annually.

Nine contamination scenarios were described in separate versions of the questionnaire. Depending on the version, the probability of future nitrate contamination ranged from only 25% to 100%, or certain contamination. In addition, the future years of expected contamination ranged from five to 40 years from the present. The latter information is related to demand uncertainty. That is, the further into the future that contamination is expected, the less likely it is that a person will demand groundwater for personal use. Nevertheless, households may have bequest values on behalf of future generations; hence, bequest attitudes were also elicited. In addition, each questionnaire requested use-related and socioeconomic data (e.g., income) from respondents. Approximately 78% of the households returned a completed questionnaire. The following results are based on these returns.

Results

Statistical methods were used to estimate the effects of income, uncertainty, and interests in future generations on household's maximum willingness-to-pay for groundwater protection. A linear version of the option price model is:

$$OP = -1064.7 + 0.0202 \times \text{Income} - 1456.5 \times \text{PrSupply}$$
$$+ 288.3 \times \text{PrDemand} + 238.5 \times \text{Bequest}$$

where OP = option price, or the household's maximum annual willingness-to-pay for groundwater protection when future supply and demand are uncertain (see Chapter 2),

Income = annual household income,

PrSupply = likelihood of future supply stated in terms of probability (For example, the probability corresponding to a 75% chance that groundwater will remain potable in the future in 0.75.),

PrDemand = a household's probability of future demand for use of Cape Cod's groundwater (also in decimal form), and

Bequest = an index from 1 (low) to 5 (high) reflecting household's attitudes about protecting groundwater on Cape Cod for use by future generations.

The results show that income, probability of future demand and concern about future generations increase maximum willingness-to-pay. In contrast, the probability of future supply decreases maximum willingness-to-pay. That is, a household's maximum willingness-to-pay decreases as the probability of future supply increases (or the threat of contamination decreases).

The statistical analysis both confirms what we expect in a qualitative way and, importantly, facilitates quantification of the economic value of groundwater by households residing in Falmouth, Massachusetts. Figure 9.1 uses histograms to illustrate the quantitative effects. The reader should recognize, however, that two-dimensional graphs such as histograms display the estimated effect of only one variable on maximum willingness-to-pay. Implicitly, then, the values of other variables are being held constant. Unless stated otherwise, the default values for these plots are the following: Income = $40,000; PrSupply = 0.25; PrDemand = 0.75, Bequest = 5. For example, the histogram that varies income is contingent on PrSupply = 0.75, PrDemand = 0.75, and Bequest = 5. Different values for these implicit variables would change the size of the bars.

The positive effect of income on willingness-to-pay is illustrated in Figure 9.1a for annual household incomes ranging from $10,000 to $100,000. Option prices corresponding to these incomes range from $182

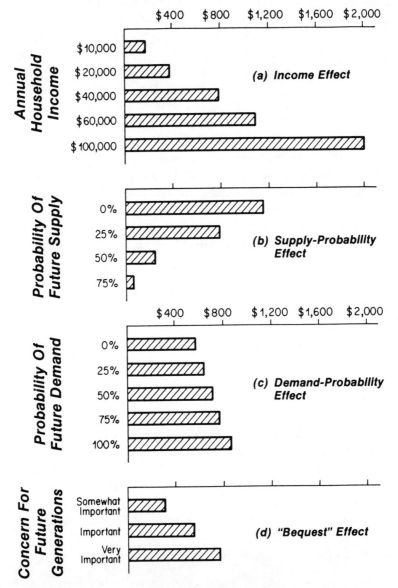

Figure 9.1 Determinants of the economic value of potable groundwater: (a) annual household income; (b) probability of future supply; (c) probability of future demand; and (d) concern for future generations.

to $2,000, increasing by about $202 for each $10,000 increase in annual income (0.0202 × $20,000 = $202). Alternatively, we can apply the elasticity formula to interpret the effect of income on maximum willingness-to-pay. In this case, we can use the coefficient on income to define the "income elasticity of willingness-to-pay" as e_{WTP} = [0.0202 × (Income/OP)] where you decide the value for an income and OP is calculated from this value and the default values for the three other variables. For example, when income = $40,000, OP = $788, and elasticity = 1.025. In other words, a 1% increase in annual household income from $40,000 is estimated to increase maximum willingness-to-pay by 1.025%. Since the elasticity value is greater than 1, the effect is elastic. Try calculating elasticities for other income levels.

Next, consider the inverse effect of the probability of future supply on maximum annual willingness-to-pay to avert future contamination (Figure 9.1b). Under the default conditions, option price decreases from $1,152 when the probability of future supply is zero (i.e., future contamination is certain) to $60 when there is only a 25% chance of future nitrate contamination. The coefficient on the PrSupply variable implies that the option price decreases by about $364 for every 25% increase in the probability of future supply (−1456.5 × 0.25 = −$364). In contrast, a 25% increase in a household's probability of future demand increases maximum annual willingness-to-pay by about $72 (Figure 9.1c).

The reader may wonder why a household with no expectation of future demand for groundwater for personal use would be willing to pay anything for protection (Figure 9.1c). One explanation is that the household may be concerned about the welfare of future generations living on Cape Cod. The effect of the bequest variable on option prices reflects this concern (Figure 9.1d). Option prices range from $311 for a household who considers groundwater protection for future generations to be "somewhat important" (Bequest = 3) to $788 for a household who considers it to be "very important" (Bequest = 5). For comparison, maximum annual willingness-to-pay ranges from $94 to $571 when the probability of future demand is zero (instead of the default value, 0.75). It appears from this single study that bequesting a potable supply of groundwater for use by future generations is a powerful motivation behind groundwater protection.

Concluding Remarks

The reader should recall the interpretation of "maximum willingness-to-pay." In economics and in the context of contingent valuation research, maximum willingness-to-pay is supposed to measure the total *net* benefits or surplus that a household/individual derives from a good or service. This study shows that misguided attempts to value groundwater at only household expenditures will grossly underestimate the total economic value of groundwater for residential use by amounts similar to those reported in Figure 9.1. Risks to health were excluded from this study because nitrate levels in public wells are monitored closely by the Commonwealth of Massachusetts. Nevertheless, the remaining net benefits, which include bequest value, option value, and damages that might occur to private property, are on par with what households in Falmouth pay in property taxes for public education. Using an average household as a reference point, maximum annual net benefits are about $125 even when the probability of future supply is high—75%. Multiplying this amount by the number of dwelling units yields a rough approximation of the total annual benefits of groundwater protection—almost $2 million. This approximation increases to $7.7 million, $13.4 million, and $19.3 million if the probabilities of future supply decrease to 50%, 25%, and 0%, respectively.

Unfortunately, common property and public good characteristics of groundwater resources prevent private markets from allocating groundwater more efficiently. However, studies such as the one reported here can be used to measure the associated non-market benefits. The next logical step is to ask whether these benefits exceed the costs of averting uncertain future contamination. Groundwater protection can be enormously expensive particularly where land use options confront high-valued coastal real estate markets. Certainly alternatives exist such as using substitute supplies (surface water or bottled water) or treating the contaminated supply. The efficient path depends on the different values at stake, the reversibility of contamination, and the probability of contamination.

Chapter 10

A Benefit-cost Analysis of Simulated Development on Cape Cod, Massachusetts

Background

Champions of environmental protection tend to make two major mistakes when arguing for restrained development in their towns. With their backs against the wall, they resort to the pseudo-economic argument that additional property tax revenue from new development will be less than the increased costs of expanding town services. This is a category mistake because it casts tax revenue as an economic benefit when it is only a transfer payment. This practice also focuses attention on the wrong issues, resulting in a mistake of omission. The obverse of the opportunity costs from providing town services is the benefits of these services—public education, police and fire protection, waste disposal, and so on. Thus, a correct analysis of the impacts of development would compare the costs and benefits of continuing to provide town services. The net effect may be zero or negative if the quality or quantity of services per capita declines. Probably more important, though, are potential non-market damages associated with groundwater, beach access, and traffic. Indeed, the potential for these damages is what actually causes alarm. Focusing on tax revenue diverts attention from the real issues.

This chapter illustrates a conceptually appropriate application of benefit-cost analysis using hypothetical development on Cape Cod, Massachusetts, as an example. Although the analysis is an exercise using hypothetical values for economic benefits and costs, it is based on realistic issues. "Best judgment forecasts" of population growth and housing development are taken from the Association for the Preservation of Cape Cod's *Growth Report* (1985) and used as a baseline for conditions without a growth management strategy. (See Devanney et al. [1976] for another, more detailed example of benefit-cost analysis applied to coastal development.)

According to the Association for the Preservation of Cape Cod, the number of houses on Cape Cod is expected to increase by about 31,000, or 27%, by the year 2000 to a total of about 144,000 (Table 10.1, Scenario A), leaving about 18% of all land that is still buildable (about 45,000 acres). Year-round population is expected to increase by about 49,000, or 30% (Table 10.1, Scenario A). In addition, seasonal population is ex-

Table

Demographic, Benefit, and Cost Data of Hypothetical De

All values are hypothetical. Monetary

	Scenario A: Data associated with APCC's					
						Year
Category	1986 (t = 0)	1987 (t = 1)	1988 (t = 2)	1989 (t = 3)	1990 (t = 4)	1991 (t = 5)
Number of houses on Cape						
Total (thousand)	115.7	118.3	121.0	123.6	126.2	128.0
Added	2620	2620	2620	2620	2620	1790
Population size on Cape						
Total peak (thous)	421.5	427.2	433.8	440.4	447.0	452.0
Added	6600	6600	6600	6600	6600	5000
Total expenditures on new houses ($million)						
	340.6	379.9	419.2	458.5	497.8	367.0
a) Profit in the construction industry ($million)						
Large companies:						
in Massachusetts	4.2	4.8	5.2	5.6	6.2	4.6
outside Mass.	4.2	4.8	.5.2	5.6	6.2	4.6
Small companies	12.8	14.5	15.7	17.2	18.7	13.8
b) Net multiplier effect ($million)						
	17.0	19.0	21.0	22.9	24.9	18.4
c) Loss in surplus due to reduced recreational opportunities ($million)						
	7.5	15.4	23.6	32.1	41.0	60.2
d) Increased costs of travel ($million)						
	4.8	9.8	15.1	20.6	26.3	30.3
e) Loss in surplus due to groundwater contamination ($million)						
	0	0	0	0	0	0
f) Increased expenses for land acquisitions ($million):						
Acquisition of land	0	0	0	0	0	0

pected to increase by 34,000 (14%), for a total peak population during the summer of 497,000 by the year 2000.

Framework

Two boundaries for external and internal benefits are evaluated: (1) all interests are internal and (2) construction companies from outside Massachusetts are external to the analysis.

10.1
velopment on Cape Cod. See text for more details.
values are not present values.

(1985) full development projections								
(time period)								
1992	1993	1994	1995	1996	1997	1998	1999	2000
(t = 6)	(t = 7)	(t = 8)	(t = 9)	(t = 10)	(t = 11)	(t = 12)	(t = 13)	(t = 14)
129.8	131.6	133.4	135.2	136.9	138.7	140.5	142.3	144.1
1790	1790	1790	1790	1790	1790	1790	1790	1790
457.0	462.0	467.0	472.0	477.0	482.0	487.0	492.0	497.0
5000	5000	5000	5000	5000	5000	5000	5000	5000
393.8	420.6	447.5	474.4	501.2	528.0	554.9	581.8	608.6
4.9	5.8	5.6	5.9	6.2	6.6	6.9	7.2	7.6
4.9	5.8	5.6	5.9	6.2	6.6	6.9	7.2	7.6
14.8	15.8	16.8	17.8	18.8	19.8	20.8	21.8	22.8
19.7	21.0	22.4	23.7	25.1	26.4	27.7	29.1	30.4
66.8	73.7	80.7	87.9	95.1	119.3	127.2	135.8	143.4
34.7	38.7	43.0	47.5	52.3	56.7	61.5	66.3	71.5
0	67.2	68.2	69.1	70.0	70.9	107.7	109.1	110.5
0	0	0	0	0	0	0	0	0

Scenario B: Data associated with a

Category	1986 (t = 0)	1987 (t = 1)	1988 (t = 2)	1989 (t = 3)	1990 (t = 4)	1991 (t = 5)
						Year
Number of houses on Cape						
Total (thousand)	115.7	117.0	118.3	119.6	120.9	121.8
Added	2594	1310	1310	1310	1310	895
Population size on Cape						
Total (thous)	420.5	424.8	428.1	431.4	434.7	437.2
Added	6500	3300	3300	3300	3300	2500
Total expenditures on new houses (millions)						
	337.2	190.0	209.6	229.2	248.9	183.5
a) Profit in the construction industry ($million)						
Large companies:						
in Massachusetts	4.2	2.4	2.6	2.7	3.1	2.3
outside Mass.	4.2	2.4	2.6	2.7	3.1	2.3
Small companies	12.6	7.1	7.9	8.5	9.2	6.9
b) Net multiplier effect ($million)						
	16.9	9.5	10.5	11.5	12.4	9.2
c) Loss in surplus due to reduced recreational opportunities ($million)						
	7.5	11.4	15.4	19.5	23.6	26.5
d) Increased costs of travel ($million)						
	4.8	7.3	9.8	12.4	15.1	17.0
e) Loss in surplus due to groundwater contamination ($million)						
	0	0	0	0	0	0
f) Increased expenses for land acquisitions ($million):						
Acquisition of land	2.0	57.0	62.9	68.8	74.6	55.1

Two alternatives are compared for each boundary condition. As indicated above, the Association's growth projections define baseline conditions, representing the option of letting property markets allocate coastal resources. This option is compared to an aggressive acquisition program whereby towns override a cap on property taxes of 2.5% and buy vacant land to prevent development.

The exercise considers the following direct and indirect effects: (1) economic profit for the construction industry; (2) net multiplier effects induced by development; (3) costs of the acquisition program; and (4) technical externalities affecting groundwater potability, traffic, outdoor recreation, and option and existence values. For convenience, it is assumed that the total net benefits of town services for residents are the same for both scenarios.

hypothetical acquisition program

(time period)								
1992	1993	1994	1995	1996	1997	1998	1999	2000
(t = 6)	(t = 7)	(t = 8)	(t = 9)	(t = 10)	(t = 11)	(t = 12)	(t = 13)	(t = 14)
122.7	123.6	124.5	125.4	126.3	127.2	128.1	129.0	129.9
895	895	895	895	895	895	895	895	895
439.7	442.2	444.7	447.2	449.7	452.2	454.7	457.2	459.7
2500	2500	2500	2500	2500	2500	2500	2500	2500
196.9	210.3	223.8	237.2	250.6	264.0	277.4	290.9	304.3
2.9	2.6	2.8	3.0	3.1	3.3	3.5	3.7	3.8
2.9	2.6	2.8	3.0	3.1	3.3	3.5	3.7	3.8
7.4	7.9	8.4	8.9	9.4	9.9	10.4	10.9	11.4
9.8	10.5	11.2	11.9	12.5	13.2	13.9	14.5	15.2
29.5	32.5	35.5	38.6	41.6	57.6	60.8	64.2	67.5
18.9	20.8	22.7	24.7	26.7	28.7	30.7	32.8	34.9
0	0	0	0	0	0	0	0	0
59.0	63.1	67.1	71.2	75.2	79.2	83.2	87.2	91.3

Since this analysis is only an exercise, the projections provided by the *Growth Report* to the year 2000 offer a convenient, albeit arbitrary, cutoff period at 15 years. Typically, the time horizon should be the life of a project or policy. Thus, the cutoff period for an acquisition program would include several generations since the effects of such a program would extend well beyond the period involving acquisition costs.

The following assumptions are made:

(1) The Association for The Prevention of Cape Cod's projections are accurate;

(2) Large construction companies build 25% of all new houses;

(3) Large construction companies (all based off Cape Cod) make a

10% economic profit, while small companies (all based on Cape Cod) make a 5% economic profit on new houses;

(4) 50% of the large construction companies are from out-of-state;

(5) The average price of a new house is $130,000 in 1985 dollars. This will grow in real (i.e., non-inflationary) terms by $15,000 per year due to increased demand and to reduced availability of buildable lots;

(6) The average price of vacant land is 30% of developed land;

(7) The net construction multiplier is 0.05 (i.e., 5% of total expenditures on new houses);

(8) With the acquisition program, towns acquire 10% of the lots that would otherwise be developed during the first year and 50% thereafter;

(9) Annual increases in travel times per household for all but recreation are one hour per 1,000 additional households (shopping, commuting to work);

(10) The cost of increased travel time is $16 per hour per household (costs of gasoline and opportunity costs of leisure time);

(11) Average surplus for outdoor recreation, option value and existence values per household per year (a) decreases by $25 per household per 1,000 people added to total peak population (decrease due to congestion and reduced access to recreational sites), (b) decreases by $100 per household when shellfishing is prohibited in most coastal lagoons, and (c) decreases by an additional $120 per household when swimming is prohibited in most lagoons and some beaches;

(12) Shellfishing and swimming are prohibited when total peak population reaches 460,000 and 480,000, respectively;

(13) The potability of wells that service 20% of the households will be lost when the number of households reaches 130,000; 30% of the households will lose their public supply of potable water when the number of households reaches 140,000;

(14) The opportunity cost per household that loses potable well water is $7 per day. This figure comprises increased costs of buying bottled water and direct losses of utility because well water is not potable.

In summary, we have delineated the boundaries of the analysis, identified the alternatives, set up a system of accounts, established the time horizon, made our assumptions explicit, and presumably conferred with experts on the likely sequence of events related to development and population growth. Next we would collect data on the benefits and costs that increase or decrease from the present for each alternative. Finally, we "crunch" the numbers to estimate net present values of baseline and acquisition alternatives under each boundary condition.

Results

The time series of economic benefits and costs for each account are listed in Table 10.1. The net benefits of either alternative are determined by subtracting the present value of loss categories c, d, e, and f from the present value of benefit categories a and b. In turn, the net present value of the status quo alternative is subtracted from that of the acquisition option to determine the economic efficiency of acquiring land to reduce growth. These final results are summarized in Table 10.2 for a range of discount rates. The results of the sensitivity analysis indicate that acquisition is an efficient alternative to unmanaged growth in this hypothetical example when the social discount rate is 4% or less.

The results illustrate how an increasing discount rate favors the influence of values early in a project. In this case, the increasing opportunity costs associated with losses in future recreational benefits, increased travel costs, and increased costs associated with potable groundwater are steadily reduced in present value terms as the discount rate increases. Con-

Table 10.2
Net Present Value (NPV) of Acquisition Program. (PV of acquisition program minus PV of full development)

Social discount rate (percent)	NPV (millions)
0	$151.8
1	108.5
2	71.0
3	38.6
4	10.6
5	−13.5
6	−34.4
7	−52.4
8	−67.8

sequently, the net present value of the acquisition program decreases with increasing discount rates (Table 10.2).

Concluding Remarks

Several points are worth making and repeating. First, although not shown, the present value of each alternative is negative. That is, non-market damages associated with growth exceed the benefits of development in each case. This result may be typical of present and future conditions since having to pay to maintain previously "free" non-market benefits automatically lowers personal utility. The advantage of the acquisition program is that it minimizes damage by controlling population growth.

Second, the 15-year time horizon does not adequately represent the temporal effects of an acquisition program since its effects would extend well beyond the period required to buy land. Thus, the present value of damages associated with recreation, travel, and groundwater is underestimated. Thirty to 50 years should be used as terminal time periods in actual applications. Given the assumptions in this exercise, the acquisition program would probably do very well at the higher discount rates as well.

Third, from a purely efficiency standpoint, all changes in economic benefits and costs should be included in a benefit-cost analysis. However, those who commission a study—those with specific geopolitical boundaries in mind—might insist on excluding specific effects. For example, it is conceivable that many would want to exclude the profits for the large construction companies that are based outside of Massachusetts. This would result in the acquisition program passing at the 5% discount rate (net present value is $15.1 million), but still not at 6% (net present value is −$7.7) and higher. However, these exclusions represent goals other than efficiency; it is the job of economists to point out such abuses of benefit-cost analysis.

Finally, these results are tied closely to the assumptions concerning the physical relationship between development and traffic and environmental quality. In practice, one should quantify these relationships before predicting effects on behavior and subsequently predicting changes in economic surpluses. This would be done in consultation with other scientists and planners. A range of probabilities for the likely values of each benefit and cost category (or at least the critical ones) can be used to determine expected values as we did in Chapter 4.

Chapter 11

Parting Remarks

The U.S. Coastal Zone Management Act of 1972 and its Amendments of 1976 admonish states to preserve, protect, develop, enhance, and, where possible, restore the resources of the coastal zone of the United States. This seemingly contradictory set of actions actually calls for a practical balance among competing interests. Although no particular approach is suggested by the Act, a satisfactory response to this admonition should be able to assess tradeoffs among competing interests in an objective, comprehensive, and theoretically sound manner. Economic analysis is one approach that satisfies these requirements and that has the added advantage of using commensurable measures for market and non-market impacts.

Market failures—particularly non-exclusive resources and externalities—abound within the world's coastal zone. There can be little doubt that the behavior of profit-maximizing firms and of utility-maximizing consumers in traditional markets often results in inefficient allocations of coastal resources. Similarly, public policies which are influenced primarily by spurious economic arguments also allocate coastal resources inefficiently.

The realization that the earth is a "spaceship"—that the natural environment has a limited capacity to assimilate the residuals of production and consumption and that our stocks of natural resources are truly scarce—has spurred economists and other scientists to consider the interactions between man and environment in great detail. The conceptual and methodological advances made in resource and environmental economics since the 1960s facilitate economic valuations of natural resources where markets do not exist. Thus, if potable groundwater, access to beaches, opportunities to go fishing, uncrowded streets, and other unpriced resources and opportunities are valued by coastal residents, there is now an opportunity to represent these interests on an equal footing with traditional economic growth. Indeed, comprehensive economic studies of coastal zone management policies may become more commonplace as people better understand and acknowledge a role for economic analysis. Consider the U.S. Supreme Court's recent ruling on the taking of property as a case in point (no pun intended). The Court ruled on 9 June 1987 that local governments violate the Fifth Amendment to the Constitution when landowners are deprived even temporarily of all use of their land. The specific case considered by the Court concerned barring the construction of buildings in a flood zone. However, the ruling is expected to have a much broader effect on resource planning since compensation for damages imposed on landowners will probably become more commonplace. In the future, effective land use policy will require an admixture of planning, legal and economic thought, and actual Pareto improvements. Ignoring the economic consequences of policy options only perpetuates poor allocations of resources.

Of course, this book only presents the necessary foundations of economic thought and analysis and certainly does not recommend specific solutions to the myriad conflicts of interest in the coastal zone. To this end, the interested reader might test his understanding of economics by considering once again (1) consumer surplus and economic profit, (2) the immense importance of opportunity costs, (3) the distinction between direct and indirect effects and the implications for net multiplier effects, (4) social discounting, (5) the pivotal role of an indifference preference structure in economic valuation, and (6) why markets and prices are not necessary for economic valuation of environmental resources. These and related concepts form a solid basis for economic inquiry from which to expand your knowledge of economic analysis.

References

Anderson, G. D., and R. C. Bishop. 1986. "The Valuation Problem." In *Natural resource economics: Policy problems and contemporary analysis,* edited by D. W. Bromley, 89–137. Boston: Kluwer Nijhoff Publishing.

Anderson, G. D., and S. F. Edwards. 1986. Protecting Rhode Island's coastal ponds: An economic assessment of downzoning. *Coastal Zone Management Journal* 14: 67–92.

Association for the Preservation of Cape Cod (APCC). 1985. *Growth report: Options for Cape Cod's future.* Orleans, MA.

Barth, M. C., and J. G. Titus, eds. 1984. *Greenhouse effect and sea level rise: A challenge for this generation.* New York: Van Nostrand Reinhold Company Inc.

Broadus, J. M., and A. G. Gaines. 1987. Coastal and marine area management in the Galápagos Islands. *Coastal Zone Management Journal* 15: 75–88.

Bromley, D. W., ed. 1986. *Natural resource economics: Policy problems and contemporary analysis.* Boston: Kluwer Nijhoff Publishing.

Devanney, J. M. III, G. Ashe, and B. Parkhurst. 1976. *Parable Beach: A primer in coastal zone economics.* Cambridge, MA: MIT Press.

Ducsik, D. W. 1974. *A handbook of social, economic, and legal considerations regarding public recreational use of the nation's coastal shoreline.* Cambridge, MA: MIT Press.

Edwards, S. F. 1984. An analysis of the nonmarket benefits of protecting salt pond water quality in southern Rhode Island: An application of the hedonic price and contingent valuation techniques. Ph.D. diss., University of Rhode Island, Kingston, R.I.

Freeman, A. M. III. 1979. *The benefits of environmental improvement: Theory and practice.* Baltimore: Johns Hopkins University Press.

Freeman, A. M. III. 1982. *Air and water pollution control: A benefit-cost assessment.* New York: John Wiley and Sons.

Hoffman, J. S. 1984. "Estimates of Future Sea Level Rise." In *Greenhouse effect and sea level rise: A challenge for this generation,* edited by M. C. Barth and J. G. Titus, 79–104. New York: Van Nostrand Reinhold Company Inc.

Hufschmidt, M. M., D. E. James, A. D. Meister, B. T. Bower, and J. A. Dixon. 1983. *Environment, natural systems and development: An economic valuation guide.* Baltimore: Johns Hopkins University Press.

Kneese, A. 1985. *Measuring the benefits of clean air and water.* Baltimore: Johns Hopkins University Press.

Levi, M. 1985. *Thinking economically: How economic principles can contribute to clear thinking.* New York: Basic Books, Inc.

Mansfield, E. 1979. *Microeconomics: Theory and applications.* 3rd edition. New York: W. W. Norton and Company.

Mishan, E. J. 1982. *Cost-benefit analysis.* 3rd edition. London: George Allen and Unwin.

Randall, A. 1981. *Resource economics: An economic approach to natural resource and environmental policy.* New York: John Wiley and Sons.

U.S. Geological Survey. 1985. *National water summary 1984: Hydrologic events, selected water quality trends, and ground-water resources.* Water Supply Paper 2275. Reston, VA.

Index

About the Author

Steven F. Edwards is presently Economist with the National Marine Fisheries Service at the Northeast Fisheries Center in Woods Hole, Massachusetts. He received a Ph.D. in resource economics from the University of Rhode Island and subsequently worked for three years as a Postdoctoral Fellow and Social Scientist in the Marine Policy Center at the Woods Hole Oceanographic Institution.